Web Recommendations Systems

K. R. Venugopal · K. C. Srikantaiah ·
Sejal Santosh Nimbhorkar

Web Recommendations Systems

Springer

K. R. Venugopal
Bangalore University
Bengaluru, Karnataka, India

K. C. Srikantaiah
Department of Computer Science
and Engineering
SJB Institute of Technology
Bengaluru, Karnataka, India

Sejal Santosh Nimbhorkar
Department of Computer Science
and Engineering
BNM Institute of Technology
Bengaluru, Karnataka, India

ISBN 978-981-15-2512-4 ISBN 978-981-15-2513-1 (eBook)
https://doi.org/10.1007/978-981-15-2513-1

This Springer imprint is published by the registered company Springer Nature Singapore Pte Ltd.
The registered company address is: 152 Beach Road, #21-01/04 Gateway East, Singapore 189721, Singapore

Foreword

This book covers the topic of Web recommendations systems and offers an overview of approaches to develop state-of-the-art systems. The authors present algorithmic approaches in the fields of Web recommendations by extracting knowledge from Web logs, Web page contents and hyperlinks. Recommendations systems have been used in diverse applications including query log mining, social networking, news recommendations and computational advertising. With the explosive and diverse growth of Web contents, Web recommendations have become a critical aspect of all search engines.

There is a growing interest in recommendations systems that suggest image, music, video, books and other products and services to users based on their likes and dislikes. Such services are important since user preferences are often complex and not readily reduced to keywords or subject categories, but rather best illustrated by an example. Hence, to enhance the user experience, the search engine provides various recommendations that are personal and of high quality. The authors discuss measures to improve the effectiveness of recommendations systems and illustrate the methods with practical case studies.

With contributions from researchers, this book strikes a balance between fundamental concepts and state-of-the-art technologies, providing readers with insights into Web recommendations systems. It is an essential reading for a broad audience, including academic researchers, research engineers and industrial practitioners due to its focus on applications and references. The readership of this book is intended to be postgraduate/doctoral researchers and professional engineers.

August 2019

N. R. Shetty
Chancellor
Central University
Kalaburgi, Bengaluru, Karnataka, India

Preface

The complexity and size of websites have grown exponentially along with the growth of the World Wide Web. Hence, it has become increasingly challenging and time consuming for the users to discover relevant information. Web recommendations systems are a subclass of knowledge refining systems that explore to predict users' preferences, used to enhance user experience on the Web and to increase the search engine usability. In recent years, they are commonly utilized in diverse fields that include search queries, Web page, music, image, movies, books, news, research articles, products and social tags and can be developed using user search log and navigation history.

Chapter 1 presents an introduction to the book. Chapter 2 addresses the problems of mining semi-structured data using cosine similarity for Web data extraction from Deep Web. Web sequential access patterns are mined to analyse cyclic behaviour and prefetching rules are generated using bi-directional growth in Chap. 3. These prefetching rules are utilized for Web caching and prefetching. Finding synonyms for a given keyword is presented in Chap. 4 while search engine result pages are ranked using probability correctness of fact. Automatic construction of Web directories using Web logs is achieved by mapping Web pages to categories using the Levenshtein similarity weight algorithm in Chap. 5.

Semantically related search queries for the given input query is illustrated by discovering keywords present in snippets clicked and un-clicked documents in feedback session in Chap. 6. Web Navigation Prediction Framework for Web page Recommendations(WNPWR) is discussed in Chap. 7. User Session Graph using user sessions from navigation log is constructed for Web page recommendations to solve *new page problem* in Chap. 8. A predicting conversion in advertising using the expectation–maximization model is discussed to provide an influence of their advertising campaigns to the advertisers by understanding hidden topics in search terms with respect to the time period in Chap. 9.

The authors appreciate the suggestions from the readers and users of this book. Kindly communicate the errors, if any, to the following email address: venugopalkr@gmail.com.

Bengaluru, India
August 2019

K. R. Venugopal
K. C. Srikantaiah
Sejal Santosh Nimbhorkar

Acknowledgements

We wish to place on record our deep debt of gratitude to Late Shri M. C. Jayadeva and Prof. K. Venkatagiri Gowda for their inspiration, encouragement and guidance throughout our lives. We thank Prof. N. R. Shetty, Former President, ISTE and Former Vice Chancellor, Bangalore University, Bangalore for his foreword to this book.

We owe debt of gratitude to Prof. L. M. Patnaik, Sri K. Narahari, Sri V. Nagaraj, Sri Mukund K. K., Prof. S. Lakshmana Reddy, Prof. K. Mallikarjuna Chetty, Prof. H. N. Shivashankar, Prof. P. Sreenivas Kumar, Prof. Kamala Krithivasan, Prof. C. Sivarama Murthy, Prof. T. Basavaraju, Prof. M. Channa Reddy, Prof. N. Srinivasan, Prof. M. Venkatachalappa, T. G. Girikumar, P. Palani, M. G. Muniyappa for their support.

We are grateful to Justice M. Rama Jois, Sri Y. K. Raghavendra Rao, Sri P. R. Ananda Rao, Sri Sreedhar Sagar, Sri Prabhakar Bhat, Prof. K. V. Acharya, Prof. Khajampadi Subramanya Bhat, Sri Dinesh Kamath, Sri D. M. Ravindra, Sri Jagadeesh Karanath, Sri N. Thippeswamy, Sri Sudhir, Sri V. Manjunath, Sri N. Dinesh Hegde, Sri Nagendra Prasad, Sri Sripad, Sri K. Thyagaraj, Smt. Savithri Venkatagiri Gowda, Smt. Karthyayini Venugopal and Smt. Rukmini T., our well-wishers for their encouragement and inspiring us to write this book.

We thank Prof. Dinesh Anvekar, Dr. P. Deepa Shenoy, Dr. K. B. Raja, Dr. K. Suresh Babu, Dr. D. N. Sujatha, Dr. Vibha L., Dr. S. H. Manjula, Dr. Thriveni J., Dr. Anita Kanavalli, Dr. Veena H. Bhatt, Dr. Sivasankari H., Dr. Shaila K., Dr. Prashanth C. R., Dr. Ramachandra, Smt. Nalini, Dr. D. Annapurna, Dr. Kumaraswamy M., Dr. Kiran K., Dr. Arunalatha J. S., Dr. Lata B. T., Dr. Krishna A. N., Prakash G. L., Gomathy Pratima, Vandana Jha, Krishna Kumar N., Srikanth P. L., Suraj M., Roopa M. S., Shailesh Kumar Gupta, Kamalakant Tiwari, Vinuth Chandrashekar, Vijay Mathapati, Rashmi V., Abhishek Datta, Thakur Ganesh Singh and Shradha G., for their suggestions and support in bringing out this book.

We express our deep gratitude and heartfelt thanks to our family members Tejaswi Venugopal, Sai Teja Vasudev, T. Shivaprakash, T. Krishnaprasad and Lakshmi Priya K., Chowdaiah, Parvatamma, Chikkaiah K. K., Honnagirigowda, Savitramma, Shashi M. H., Shreya K. S., Tanaya K. S., Jayantilal, Taralaben, Chintaman, Nalini, Santosh N., Jeet, for all their support, patience and encouragement.

Contents

About the Authors

Dr. K. R. Venugopal is the Vice Chancellor of Bangalore University. He holds eleven degrees, including a Ph.D. in Computer Science Engineering from IIT-Madras, Chennai and a Ph.D. in Economics from Bangalore University. He also has degrees in Law, Mass Communication, Electronics, Economics, Business Finance, Computer Science, Public Relations and Industrial Relations. He has authored and edited 68 books and published more than 800 papers in refereed international journals and international conferences. Dr. Venugopal was a post-doctoral research scholar at the University of Southern California, USA. He has been conferred with IEEE fellow and ACM Distinguished Educator for his contributions to computer science engineering and electrical engineering education.

Dr. K. C. Srikantaiah is a Professor at the Department of Computer Science and Engineering at SJB Institute of Technology, Bangalore, India. He received his B.E. from Bangalore Institute of Technology, M.E. from University Visvesvaraya College of Engineering, Bangalore, in 2002 and Ph.D. degree in Computer Science and Engineering from Bangalore University in 2014. He has published 20 research papers and authored a book on Web mining algorithms. His research interests include data mining, Web mining, big data analytics, cloud analytics and the Semantic Web.

Dr. Sejal Santosh Nimbhorkar is an Associate Professor at B N M Institute of Technology. She has more than 15 years of industry, research and teaching experience. She holds M.E. and B.E. degrees in Computer Science and Engineering from University Visvesvaraya College of Engineering and Gujarat University, respectively. She has published 18 papers in refereed international journals and international conferences. She received an outstanding paper award at the 2015 European Conference on Data Mining. Dr. Nimbhorkar has also received project grants from Karnataka State Council for Science and Technology (KSCST). Her research interests include mining, Web mining, sentiment analysis and IoT.

Acronyms

AAbMC	Aggregate Absorbing Markov Chain
ABIR	Annotation-Based Image Retrieval
ACSIR	ANOVA Cosine Similarity Image Recommendation
AMC	Aggregate Markov Chain
ANN	Artificial Neural Network
ANOVA	Analysis of Variance
AOL	America On-Line
ARM	Association Rule Mining
CBD	Clustering by Direction
CBIR	Content-Based Image Retrieval
CC	Color Correlogram
CG	Cover Graph
CLF	Common Log Format
CM	Color Moments
CMA	Cyclic Model Analysis
CPC	Cost Per Click
CPM	Conceptual Prediction Model
CTR	Click Through Rate
DAG	Directed Acyclic Graph
DDC	Dewey Decimal Classification
DIM	Document Influence Model
DOM	Document Object Model
DTM	Dynamic Topic Model
EC	Example Classifier
ECLF	Extended Common Log Format
ED	Edge Detection
EM	Expectation Maximization
FED	Frequent Episode discovery
GBDT	Gradient Boosted Decision Tree
HAL	Hyperspace Analogue to Language

Hdiff	Heat Diffusion
HDM	Heuristic-based Distributed Miner
HNB	Hidden Nave Bayes
HTML	HyperText Mark up Language
IE	Information Extraction
IIS	Internet Information Server
IR_URFS_VF	Image Recommendation with User Relevance Feedback Session and Visual Features
IRAbMC	Image Recommendation with Absorbing Markov Chain
KDD	Knowledge Discovery from Data
LARS	Location-Aware Recommendation System
LASSO	Least Absolute Shrinkage and Selection Operator
LBP	Local Binary Pattern
LCC	Library of Congress Classification
LD	Levenshtein Distance
LDA	Latent Dirichlet Allocation
LSA	Latent Semantic Analysis
LSW	Levenshtein Similarity Weight
MAE	Mean Absolute Error
MAP	Mean Average Precision
MFP	Maximum Forward Path
MI	Mutual Information
MRR	Mean Reciprocal Rank
MSI	Markovian Semantic Indexing
NGD	Normalized Google Distance
ODP	Open Directory Project
PAM	Pachinko Allocation Model
PCA	Principal Component Analysis
PCAEM	Prediction Conversion in Advertising using Expectation Maximization
PD	Pseudo Document
PLSA	Probabilistic Latent Semantic Analysis
PMI	Pointwise Mutual Information
QC	Query Click
QFG	Query Flow Graph
QH	Query Hashing
QRG	Query Relevance Graph
QRGQR	Query Relevance Graph for Query Recommendation
RF	Relevance Feedback
RMSE	Root Mean Square Error
RSR	Related Search Recommendation
RWR	Random Walk with Restart
SCM	Snippet Click Model
SEM	Search Engine Market
SERPs	Search Engine Result Pages

SIG	Special Interest Group
SMAP	Sequential Mobile Access Pattern
SMT	Statistical Machine Translation
SOM	Self Organizing Maps
SPC	Single-Pass Clustering
SVM	Support Vector Machine
TF	Term Frequency
TF-IDF	Term Frequency–Inverse Document Frequency
TPM	Transition Probability Matrix
TPV	Topic Proportion Vector
TX	Texture
UOFS	University of Saskatchewan
URFS	User Relevance Feedback Session
URI	Uniform Resource Identifier
URL	Uniform Resource Locator
USG	User Session directed Graph
WAS	Web Access Sequence
WNPWR	Web Navigation Prediction Framework for Web Page Recommendation
WRUSG	Web Page Recommendations based on User Session Graph
WWW	World Wide Web

Chapter 1
Introduction

K. R. Venugopal, K. C. Srikantaiah and Sejal Santosh Nimbhorkar

1.1 World Wide Web

The Internet [1] is a communication network of computers forming a global system. The Internet has stimulated the adaption of website technology for publishing, blogging and Web feeds. With the advent of the Internet, new types of applications for interaction such as instant messaging and Social Networking are developed [2]. The power of the Internet has lead to taking business online with the development of e-commerce Web applications (i.e. business-to-customer, business-to-business and financial services) impacting across entire industries. Coming to the point of governance, it has no centralized technology or policies for access and usage. Each individual network connected to the Internet has its own policies and access controls [2].

The interconnected computing systems across the world form the Web [3, 4]. Web is also known as a collection of resources and information interconnected *via* the Internet. With the evolution of the Internet, the major challenges were: (i) how to establish a network of users such as researchers, who wish to share their research work with other researchers to get reviews and suggestion (ii) how to represent the data to be shared and finally how to access the shared data. The establishment of the network of researchers could happen with the development of Web applications. The challenge regarding the access of data was solved with the development of Hyper

K. R. Venugopal (✉)
Bangalore University, Jnana Bharathi, Bengaluru 560056, India
e-mail: venugopalkr@gmail.com

K. C. Srikantaiah
SJB Institute of Technology, BGS Health and Education City, Uttarahalli Main Road, Kengeri, Bengaluru 560060, India
e-mail: srikantaiahkc@gmail.com

S. Santosh Nimbhorkar
BNM Institute of Technology, Banashankari, Bengaluru 560060, India
e-mail: sej_nim@yahoo.co.in

K. R. Venugopal et al., *Web Recommendations Systems*,
https://doi.org/10.1007/978-981-15-2513-1_1

Text Transfer Protocol (http) [5] and representation of information to be shared was solved with the advent of Hypertext Mark up Language (HTML).

It has become common in everyday life to refer to the term Internet or Web, when using a Web browser to see Web pages of interest. Often the term the Internet and Web are mistaken; Web is one of the services offered by the Internet. The World Wide Web (Web) [4, 6] is collection of heterogeneous content such as Web pages (interconnected documents) containing html, XML, JSON objects and other multimedia objects referenced by Uniform Resource Identifier (URI).

Search engine is a Web tool that accepts query keywords as inputs, searches the keywords in its database and provides the pages that contain these keywords as Search Engine Result Pages (SERPs); it operates in the order: *Web Crawling, Indexing, Searching and Ranking*. Web search engines are classified into the categories: *Crawler based, Topic Directories, Hybrid Engines, Meta Engines and Specialty Search Engines*.

Exponential growth of Web information had made it a challenging task for the search engines to meet the users' requirements. Handling Web information appropriately and organizing adequately is more demanding on the Web. To get any information from Web, the user issue queries, follows some links in Web snippets, clicks on advertisements and spends some time on pages. The user reformulates his query, if he is not convinced with the clicked page information. In order to enhance the user experience, the search engine provides various kind of recommendations like queries, Web pages and images.

1.2 Web Mining

The increase in the number of Web users and size of the data generated indirectly poses a challenge in satisfying the Web user needs in terms of their preferences, security, response time, etc. Hence, it is necessary to analyse the behaviour of the Web user from their browsing history to identify the user preferences, that help in improving the business process. There is a need for Data Mining to derive business rules from user transaction data.

Data Mining is the process of extracting useful information from a large repository of data. It is also referred to as Knowledge Discovery from Data (KDD). Alternatively, it can also be viewed as an essential step in the process of knowledge discovery which consists of an iterative sequence of: (i) Data Preprocessing, (ii) Data Mining and (iii) Data Post-processing. Data Mining is an integration of various fields such as Database Technology, Statistics, Machine Learning, Information Science, Visualization, etc.

The Data Mining functionalities determine the kind of patterns that can be mined from the given data and they consist of: *Characterization and Discrimination*, *Mining Frequent Patterns, Association and Correlations*, *Classification*, *Clustering*, *Outlier Analysis*, *Evolution Analysis*.

Data Mining can be used to perform trend analysis and predict future behaviour in various applications. Hence, it allows businesses to make proactive, helps in taking

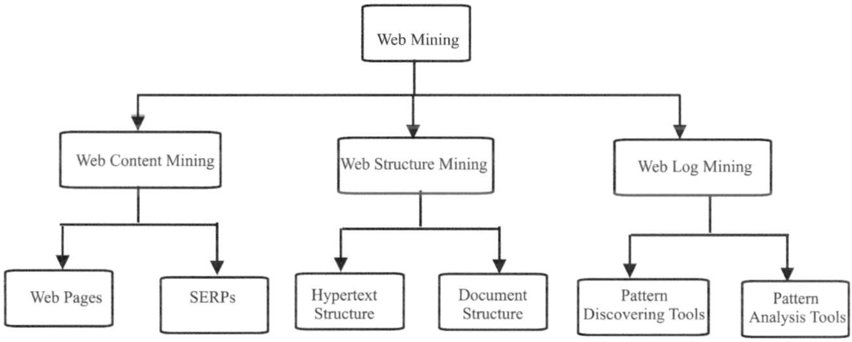

Fig. 1.1 Classification of Web mining

knowledge-driven decisions and can also answer business questions that traditionally were too time consuming to resolve, etc. It can be classified based on various criteria such as data models, types of data, applications involved, techniques utilized, etc. If the classification is based on the type of data, then we have Text Mining for text document database, Spatial Data Mining for spatial database, Time-series Mining for time-series database, Multimedia Data Mining for multimedia database, Web Mining for mining on Web data, etc.

Web mining intends to determine valuable knowledge and information from logs, Web page contents and URLs. Web mining can be broadly classified into three categories based on type of Web data: Web Content Mining, Web Usage(log) Mining and Web Structure Mining. Web Content Mining finds valuable knowledge and information from contents of Web pages. Web Usage Mining is the process of extracting information such as user behaviour from Web logs. Web Structure Mining determines information from links of Web pages.

The taxonomy of Web mining is shown in Fig. 1.1.

The main purpose of Web content mining is to analyse the Web pages in order to find information of interest or relevance embedded on the Web page. Web content mining can be broadly classified into two subdivisions. The first one is information retrieval and its purpose is to find useful information for locating relevant Web pages in a large collection (Web Searching). The refinement of the query is done by analysing the Web content [7]. The information embedded in the Web page is unstructured or semi-structured. Extracting information from the unstructured and semi-structured format poses greater challenges. The second subdivision is based on information extraction and its purpose is to find the structural information, which is saved in the database and to process it accordingly.

Web usage mining is also called as Web log mining, which is used to analyse the behaviour of online users [8]. This has led to two types of tracking; one is general access tracking and another, customize usage tracking [9]. The general access tracking is used to predict the customer behaviour on the Web. The Web log is located in three different locations such as Web server log, Web proxy server and

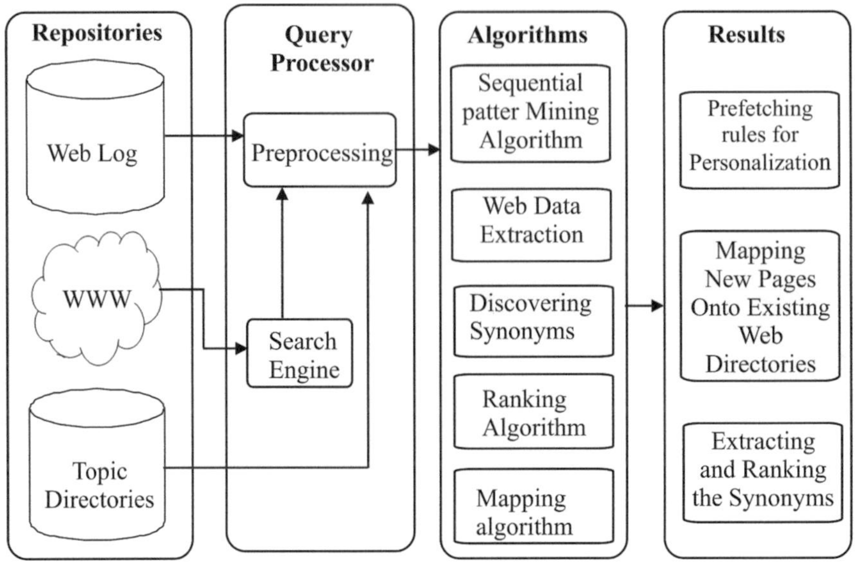

Fig. 1.2 Architecture of Web log mining and semi-structured data mining

client browser. Web log contains a large amount of irrelevant data such as noisy data, incomplete data, eroded and unnecessary information. Web server log files are used to identify the errors and failed requests, given by the webmaster or the system administrator.

Web Structure Mining is the process to analyse a Web graph, where the nodes represent Web pages and the edges represent the hyperlinks among them using graph theory. According to the type of Web structural data, Web structure mining can be divided into two categories: (i) extracting patterns from hyperlinks in the Web and (ii) mining the document structure, i.e. analysis of the DOM tree structure of the pages to describe HTML or XML tag usage.

Architecture of the proposed Web log mining and semi-structured data mining is shown in Fig. 1.2.

1.2.1 Issues in Web Mining

There are a number of research issues relevant to Web Mining and the most prominent among them are the improvement in the quality of keyword-based searches and ranking their results, effective extraction from deep Web, effective Web log mining and analysis tools, automatic construction of topic directories (such as Open Directory Project (ODP), Yahoo! directory), effectual automatic classifiers and clustering mechanisms.

Web Mining alleviates these problems by contributing solutions such as (i) mining SERPs, (ii) hyperlink analysis for ranking, (iii) automatic classification of Web documents, (iv) mining patterns from Web logs, (v) clustering of Web pages, (vi) construction of a multidimensional, multilevel Web and (vii) finding structure of the Web page. These Web Mining tasks can be used for applications like ranking, solving polysemy problem and finding user browsing patterns for customization, personalization and generation of prefetching rules.

1.3 Web Recommendations

Decades ago in a small town, everybody knew one another. Knowing each other, one can understand the interest of other and respond accordingly. The responses were usually matching the preferences of the other person. For example, when a woman visits a textile shop, the salesperson would recommend material according to the woman's preference. The salesperson understood the interests of most of the population in the town from their shopping history. So life in smalltown was all about social connections.

A few years later, the rate of interactions has gone down or the growth in the population might have made an impact. The salesperson in a book store usually can recommend new arrivals for a regular customer, but not for anybody from the town. Presently, the towns have become cities and small departmental stores have changed to supermarkets with a large number of choices. Hence, the problem arises in understanding customer preferences.

The trend changed with Amazon introducing the retail business online in the twenty-first century. If one wants to buy a book, Amazon has more than 2 million choices. Similarly, Netflix has more than 1,00,000 titles and *iTunes* has billions of tracks. With the diversity in choices, the problem is of finding the relevant item. The problem is getting worse as every minute terabytes of media are added on the Internet, hours of videos are getting uploaded on YouTube, and hundreds of books are published every hour. It gets more and more difficult to find the relevant stuff in the depth of possibilities. There is a necessity for some computational help to find items among the billions of choices. There is a need to explore methods combining the people's interest and disinterest to mine relevant items.

Recommendation technique which has branched out from data mining is not only about selling products and increasing profit; these methods of recommendation are also applied in politics to identify weaknesses to gain votes and police personnel to identify the trouble makers and terrorists.

Recommendation methods are focused on finding patterns in data. If the data size is small, anyone can become an expert to find the patterns in data using mental models. Recommendation enables the ability to handle a large quantity of information to make predictions.

1.4 Classification of Recommender System

1.4.1 Query Recommendations

Exponential gain of Web information is a challenging task for the search engines to meet the users' requirements. Handling Web information appropriately and organizing adequately is more demanding on the Web. To get any information from Web, the user issues queries, follows some links in the Web snippets, clicks on the advertisement and spends some time on pages. The user reformulates his query, if he is not convinced with the clicked page information. In order to enhance the user experience, the search engine provides various kinds of query recommendations.

Query recommendations is a technique to recommend related queries to users' input query by finding an association between queries from users' search log. It is an efficient way to enhance keyword-based search that is extensively useful to Web search systems. Users need to modify queries often because queries are informational. Users may seek discrete information on a distinct subject, hence they may check various query terms. Users may not have sufficient knowledge on a topic, and therefore adequate terms are not known to retrieve the required information.

Kato et al. [10] observed that query recommendations are frequently used when (1) an initial query is an exceptional query, (2) a single-term query is used as input query, (3) explicit queries are suggested, (4) suggestions are provided based on the modification of an input query, (5) various URLs have been clicked by users on the resulting search page. Query recommendations provided to the user efficiently can reduce the complexity of the search and help them to locate the required information more precisely. This method is extensively accepted by product, music, video search, retrieval of medical information and patient search information. Query suggestions techniques are implemented by commercial search, such as *Searches related* in Google, *Search Assist* in Yahoo!, and *Related Searches* in Bing Search.

Challenges in Query Recommendations: There are several challenges in designing query recommendations framework on the Web. First, usually users' submit short queries between one and three terms and are generally ambiguous. We observe from America On-Line (AOL) [11] data that 9.82% of Web queries contain one term, 27.31% of Web queries contain two terms and 26.99% of Web queries contain three terms. Second, in most of the cases, the users do not have sufficient knowledge of the topic that is searched, and they are not able to clearly phrase the query words. Then, users have to rephrase the query words and rephrase their queries frequently. Hence, it is necessary to solve query recommendation to satisfy users' information needs and to increase the search engine usability. Various kinds of data are used for suggestions and these data can be converted to graphs and can be used to solve many suggestions problems by designing a generic graph recommendation approach.

1.4.2 Web Page Recommendations

Internet usage has increased excessively as a result of evolution in *e*-commerce, research, *e*-banking, education, news, music, movies and electronic devices. Hence, a huge amount of information is archived and it keeps growing rapidly without any control. The decreasing costs in secondary storage have made it easy to store a huge volume of information. The valuable information is beneficial to determine interesting and useful patterns that are used by many researchers for guiding the users to visit the Web pages during their activity on the Web. This type of system is called Web Page Recommender System and helps to predict the user request.

The Web users generally spend most of the time on authoring and browsing than on search. Hence, search engines cannot efficiently predict users' search objective. Web prediction is performed on the navigation history as it is conducted in a passive manner. Hence, the prediction is performed only after the user submits his/her queries to search engines. The top k users are involved in a similar activity in Web prediction and it can be used in the recommendations system.

The likelihood of visiting a Web page by a user depended on the history of previously accessed Web pages is known as Web prediction. The Web users' prediction behaviour is trivial in Web mining to improve the search engine performance. The Web is configured as a graph, where each node represents a Web portal, and the edge represents the users' navigation. The user-visited Web pages distribution can be calculated and utilized in re-ranking and re-weighting results. The navigation path information is of prime significance than the user query. Storage of predicted Web pages in the cache can improve performance of search engine.

Behavioural targeting is a prime concern of predicting Web users' future behaviour. Behavioural targeting is an approach for efficiency improvement in advertising by online website advertisers and publishers by extracting knowledge of users' web-browsing practices. Behaviour targeting publishes advertisement through users' web-browsing actions. The analysis of user behaviour on the Web is the point of interest in on-line advertising and accurately targeted advertisements help in generating more consumer interest.

Web users' shopping style prediction has a crucial role in product recommendations, i.e. dynamic shopping recommendations across mobile, email and Web channels. It depends on each customer's current and past purchases practices. It also helps to improve conversions, website optimization and increase revenue by making related product recommendations to the customers.

Challenges in Web Page Recommendations: The World Wide Web (WWW) has generated immense opportunities to extract and gain massive online information. This inspires researchers to learn the navigation behaviour of Web site visitors from Web usage data to reduce access latency, Web page recommendations using efficient Web prediction technique and to improve the quality of service of that site. The Web log records the users' navigational behaviour. The preprocessing of the raw data is required before giving the data as input to prediction model. The preprocessing challenges include handling a huge amount of data, obtaining domain intelligence and session identification. Expensive training and low accuracy and are fundamental issues in prediction.

1.4.3 Image Recommendations

Billions of images are available on the Internet with the development of the World Wide Web. It is difficult for users to access and find image of their interest as the number of digital images has grown tremendously on the Web. Hence, additional processing is necessary to retrieve relevant images as per the user requirement. An image retrieval system provides an effective way to retrieve a set of images to meet the users' demand.

There are two basic image retrieval techniques: (i) Content-Based Image Retrieval (CBIR) and (ii) Annotation-Based Image Retrieval (ABIR). In the CBIR technique, images are retrieved based on texture, colour and shape features or by extracting knowledge of image rather than the metadata associated with the image such as tags, keywords or descriptions. The semantic meaning of the user query and low-level visual features of images do not match in CBIR. In CBIR techniques, search results are refined continuously by using the relevant feedback of user. This method is impractical for a very large dataset as it requires intensive computation.

Vertical search engine is used to perform domain-specific search and provides actual content (product) rather than links. Content-based and text-based search approaches are extensively used for these types of search engines to retrieve images. The text-based image search is dependent on the occurrence of input query terms either in surrounding text or metadata of images. This approach is widely used as it requires lower computation cost and provides faster response. Whereas, it fails to retrieve the images that are relevant but do not have the term in the surrounding text. Visual features of query-image are compared with images present in database in the content-based image search. It captures images that are relevant irrespective of the query-term as it performs content-based matching. This method requires higher time complexity and has a slower response time.

Sometimes, search engines fail to retrieve information as per the user wish because of various reasons: (i) improper input query (ii) lack of users' understanding about the input search query (iii) wrongly tagged images present in database. Hence, users are unable to obtain the desired output. This gap between users' search intention and understanding of the objects is called as semantic gap and is common in most of the image search engines.

Challenges in Image Recommendations: Web image search engines like Google and Yahoo! retrieve images with text-based queries. These text queries are matched with textual information such as comments, tags, surrounding text, URLs and titles along with Web images. Currently, only 10% of Web images have a meaningful description (annotation). Although the search engine retrieves images efficiently, they are able to maintain around only 42% precision and 12% recall [12]. Searches do not find relevant results on Google search for 52% of 20,000 queries [13]. This is an account of two main reasons: (i) generally, queries are short and ambiguous, e.g. the query *DM* has two different meaning *Data Mining* and *Data Mart*, and (ii) users may have different perspective for the same query, e.g. for query *apple*, users with apple product have different meaning than users who like apple fruit. Therefore, it is necessary to improve image recommendations results in order to satisfy users' needs and usability of the search engine.

References

1. https://en.wikipedia.org/wiki/Internet
2. http://oer.nios.ac.in/wiki/oer/ictapplication/internetanditsusage/internet_applications_and_services.html
3. http://www.internetsociety.org/internet/what-internet/history-internet/brief-history-internet
4. https://en.wikipedia.org/wiki/World_Wide_Web
5. https://en.wikipedia.org/wiki/Hypertext_Transfer_Protocol
6. http://www.slideshare.net/karthikanadar/world-wide-web-26195249
7. H. Han, T. Noro, T. Tokuda, An automatic web news article contents extraction system based on RSS feeds. J. Web Eng. **8**(3), 268284 (2009)
8. A. Ranade, A.R. Joshi, Techniques for understanding user usage behavior on the internet. Int. J. Comput. Appl. (09758887) **92**(7) (2014)
9. K. Bhalla, D. Prasad, Data Preparation and Pattern Discovery For Web Usage Mining
10. M.P. Kato, T. Sakai, K. Tanaka, When do people use query suggestion? A query suggestion log analysis. J. Inf. Retr. **16**(6), 725–746 (2013)
11. G. Pass, A. Chowdhury, C. Torgeson, A picture of search, in *Proceedings of First International Conference on Scalable Information Systems* (2006)
12. K. Stevenson, C. Leung, Comparative evaluation of web image search engines for multimedia applications, in *Proceedings of IEEE International Conference on Multimedia and Expo, ICME 2005* (2005), pp. 4–14
13. B. Smyth, A community-based approach to personalizing web search. IEEE J. Comput. **40**(8), 42–50 (2007)

Chapter 2
Web Data Extraction and Integration System for Search Engine Results

K. R. Venugopal and K. C. Srikantaiah

Abstract There is an explosive growth of information in the World Wide Web thus posing a challenge to Web users to extract essential knowledge from the Web. Search engines help us to narrow down the search in the form of Search Engine Result Pages (SERP). Web Content Mining is one of the techniques that help users to extract useful information from these SERPs. In this chapter, we propose two similarity-based mechanisms; Web Data Extraction using Similarity Function (WDES), to extract desired SERPs and store them in the local depository for offline browsing and Web Data Integration using Cosine Similarity (WDICS), to integrate the requested contents and enable the user to perform the intended analysis and extract the desired information. Our experimental results show that WDES and WDICS outperform Data Extraction based Partial Tree Alignment (DEPTA) [1] in terms of Precision and Recall.

2.1 Introduction

The World Wide Web (WWW) has now become the largest knowledge base in human history. The Web encourages decentralized authorizing in which users can create or modify documents locally, which makes information publishing more convenient and faster than ever. Because of these characteristics, the Internet has grown rapidly, which creates a new and huge media for information sharing and exchange. Most of the information on the Internet cannot be directly accessed *via* the static link,

K. R. Venugopal (✉)
Bangalore University, Jnana Bharathi, Bengaluru 560056, India
e-mail: venugopalkr@gmail.com

K. C. Srikantaiah
SJB Institute of Technology, BGS Health and Education City, Uttarahalli Main Road, Kengeri, Bengaluru 560060, India
e-mail: srikantaiahkc@gmail.com

K. R. Venugopal et al., *Web Recommendations Systems*,
https://doi.org/10.1007/978-981-15-2513-1_2

must use Keywords and Search Engine. Web search engines are programs used to search information on the WWW and FTP servers that check the accuracy of the data automatically. When searching for a topic in the WWW, it returns many links or Web sites related, i.e. Search Engine Result Pages (SERP) on the browser to a given topic. Some data on the Internet that is visible to the search engine is called surface Web, whereas some data such as dynamic data in dynamic database invisible to the search engine is called deep Web.

There are situations in which the user needs those Web pages on the Internet to be available offline for convenience. The reason being offline availability of data, limited download slots, storing data for future use, etc. This essentially leads to downloading raw data from the Web pages on the Internet, which is a major set of the inputs to a variety of software that are available today for the purpose of data mining. Web data extraction is the process of extracting the information that users are interested in, from semi-structured or unstructured Web pages and saving the information as the XML document. During Web data extraction phase, search engine result pages are crawled and stored in the local repository. Web database integration is a process of extracting the required data from the Web pages stored in the local repository and integration of the extracted data that needs to be stored in the database.

In recent years, there have been rapid improvements on technology with products differing in the slightest of terms. Every product needs to be tested thoroughly and the Internet plays a vital role in gathering information for the effective analysis of the products. In our approach, we replicate search engine result pages locally based on comparing page URLs with a predefined threshold. The replication is such that the pages are accessible locally in the same manner as on the Web. In order to make the data available locally to the user for analysis, we extract and integrate the data based on the prerequisites which are defined in the configuration file.

In a given set of Web pages, it is difficult to extract matching data. So, we have to develop a tool that is capable of extracting the exact data from the Web pages. In this chapter, we have developed the WDES algorithm, which provides offline browsing of the pages. Here, we integrate the downloaded content onto a defined database and provide a platform for efficient mining of the data required.

2.2 Related Works

Zhai et al. [1] propose Data Extraction based Partial Tree Alignment (DEPTA) algorithm for the structured data extraction from the Web based on partial tree alignment, and structured data extraction from arbitrary Web pages. The main objective is to automatically segment data records in a page, extract data items/fields from these records and store the extracted data in a database. It consists of two steps, i.e. (i) identifying individual records in a page using visual information and (ii) aligning and extracting data items from the identified records using tree matching and a novel partial alignment technique, respectively.

Ananthanarayanan et al. [2] propose a method for offline Web browsing that is minimally dependent on real-time network availability. The approach makes use of the Really Simple Syndication (RSS) feeds from Web servers and prefetches all new content specified, defining the content section of the home page. It features intelligent prefetching, robust and resilient measures for intermittent network handling, template identifier and local stitching of the dynamic content into the template. It does not provide all the information on the page. The drawback is, the content defined in the RSS feeds may not be updated nor they do provide for dynamic changes in the page.

Myllymaki et al. [3] describe ANDES, a software framework that provides a platform for building a production-quality Web data extraction process. The key aspects are that it uses XML technology for data extraction, including XHTML and XSLT and provides access to the deep Web. It addresses the issues of website navigation, data extraction, structure synthesis, data mapping and data integration. The framework shows that production-quality Web data extraction is quite feasible and that incorporating domain knowledge into the data extraction process can be effective in ensuring the high quality of extracted data. Data validation technique, a cross-system support, and a well-established framework has not been addressed.

Yang et al. [4] propose a novel model to extract data from deep Web pages. It consists of four layers, among which, the access schedule, extraction layer and data cleaner are based on the rules of structure, logic and application. The model first uses the two regularities of the domain knowledge and interface similarity to assign the tasks that are proposed from the users and chooses the most effective set of sites to visit. Then, the model extracts and corrects the data based on the logical rules and structural characteristics of the acquired pages. Finally, it cleans and orders the raw dataset to adapt to the customs of the application layer for later use by its interface.

Yin et al. [5] propose Web page templates and DOM technology to effectively extract simple structured information from the Web. The main contents include the methods based on edit distance, DOM document similarity judgement, clustering methods of Web page templates and programming an information extraction engine. The method provides steps for information extraction using DOM tree parsing and on how a page similarity judgement is to be made. The template extraction and reconstruction is depicted in order to find out how the data has been parsed on the Web page and in reconstructing the page to overcome the noise on the page. It does not parse through the dynamic content of scripts on the page.

Liu et al. [6] propose a method to extract the main text from the body of a page based on the position of DIV. It reconstructs and analyses DIV in a Web page by completely simulating its display in a browser, without additional information such as Web page template; its implementation complexity is quite low. The core idea includes the concept of atomic DIV, i.e. a DIV block that does not include other DIVs. Then it filters out redundancy and reconstructs by clustering, reposition analysis and finally stores the elements of the data in an array. The selected DIVs in the selected array contains the main text of the page. We can get the main text by combining these DIVs. The method has high versatility and accuracy. The majority of the Web page content has been made up of tables and this approach does not address the table layout data. This drastically reduces the accuracy of the entire system.

Dalvi et al. [7] explore a novel approach based on temporal snapshots of Web pages to develop a tree-edit model of HTML to improve wrapper construction. The model is attractive as a source tree transformed into a target tree can be estimated efficiently, in quadratic time, making it a potentially useful tool for a variety of tree-evolution problems. An improvement in the robustness and performance and ways to prune the trees without hampering model quality is to be dealt with.

Novotny et al. [8] represent a chain of techniques for the extraction of object attribute data from the Web pages. They discover data regions containing multiple data records and also provide a detailed algorithm for detail page extraction based on the comparison of two html subtrees. They describe the extraction from two kinds of Web pages: master pages containing structured data about multiple objects and detail pages containing data about single product, respectively. They combine the techniques of the master page extraction algorithm, detail page extraction algorithm and comparison of sources of two Web pages. The approach makes use of attribute values based on the Document Object Model (DOM) structure described in the Web pages. It has better precision of extraction of values from pages defined in the master and detailed format and also minimizes the user effort. Enhancements to the approach may include the page-level complexity of having multiple interleaved detail pages to be traversed, coagulation of different segments in the master page and series implementation of pages.

Nie et al. [9] provide an approach for obtaining data from the result set pages by crawling through the pages for data extraction based on the classification of the URL (Unified Resource Locator). It extracts data from the pages by computing the similarity between the URLs of hyperlinks and classifying them into four categories, where each category maps into a set of similar Web pages, which separates result pages from others. It makes use of the page probing method to verify the correctness of classification and improves the accuracy of crawled pages. The approach makes use of the minimum edit distance algorithm and URL-field algorithm to calculate the similarity between URLs of hyperlinks, respectively. However, there are a few constraints to this approach. It is not able to resolve issues of pages related to the partial page refreshments by the use of JavaScript engines.

Papadakis et al. [10] describe STAVIES, a novel system for information extraction from Web data source through automatic wrapper generation using clustering technique. The approach is to exploit the format of the Web pages to discover the underlying structure in order to finally infer and extract pieces of information from the Web page. The system can operate without any human intervention and does not require any training.

Chang et al. [11] have studied the major Web data extraction systems and compare them in three dimensions: the task domain, the automation degree and the techniques used. These approaches emphasize on availability of robust, flexible Information Extraction (IE) systems that transform the Web pages into program-friendly structures such as a relational database. It mainly focuses on the IE from semi-structured documents and discusses only those that have been used for Web data. The trend of highly automatic IE systems saves the effort of programming, labelling, enhancements for applying the techniques to non-html documents such as medical records and curriculum vitae and facilitates the maintenance of larger semi-structured documents.

Angkawattanawit et al. [12] propose an algorithm to improve harvest rate by utilizing several databases like seed URLs, topic keywords and URL relevance predictors that are built from previous crawl logs. Seed URLs are computed using BHITS [13] algorithm on previously found pages by selecting pages with high hub and authority scores that are used for future recrawls. The interested keywords for the target topic are extracted from the anchor tags and title of previously found relevant pages. Link crawl priority is computed as a weighted combination of popularity of the source page, similarity of link anchor text to topic keywords and predicted link score are based on previously seen relevance for that specific URL.

Aggarwal et al. [14] propose the concept of intelligent crawling where the user can specify an arbitrary predicate such as keywords, document similarity, etc., which are used to determine documents relevance to the crawl; the system adapts itself in order to maximize the harvest rate. A probabilistic model for URL priority prediction is trained using URL tokens, information about content of in-linking pages, number of sibling pages matching the predicate and short-range locality information

Chakrabarti et al. [15] propose models for finding URL visit priorities and page relevance. The model for URL ranking called apprentice is trained online by samples consisting of source page features and the relevance of the target page but the model for evaluating page relevance can be anything that outputs a binary classification. For each retrieved page, the apprentice is trained based on information from the baseline classifier and features around the link extracted from the parent page to predict the relevance of the page pointed to by the link. Those predictions are then used to order URLs in the crawl priority queue. The number of false positives has decreased significantly.

Ehrig et al. [16] propose an ontology-based algorithm for page relevance computation which is used for Web data extraction. After preprocessing, words occurring in the ontology are extracted from the page and counted. The relevance of the page with regards to user-selected entities of interest is then computed by using several measures such as direct match, taxonomic and more complex relationships on ontology graph. The harvest rate of this approach is better than baseline-focused crawler.

Srikantaiah et al. [17] propose an algorithm for Web data extraction and integration based on URL similarity and cosine Similarity. The extraction algorithm is used to crawl the relevant pages and store them in a local repository. The integration algorithm is used to integrate similar data in various records based on cosine similarity.

2.3 System Architecture

2.3.1 Problem Definition

Given a start page URL and a configuration file, the main objective is to extract pages which are hyperlinked from the start page and integrate the required data for analysis using data mining techniques. The user has sufficient space on the machine to store the data that is downloaded. The basic notations used in the model are shown in Table 2.1.

The entire system has been divided into three blocks that are able to accomplish the dedicated tasks, working independently from one another. The first block, namely the Web Data Extractor connects to the Internet to extract the Web pages described by the user and stores it onto a local repository. It also stores the data in a format that is likely to be an offline browsing system. Offline Browsing means that the user is able to browse through the pages that have been downloaded, by the use of a Web browser without having to connect to the Internet. Thus, it would be convenient for the user to go through the pages as and when he needs it.

The second block, namely the Database Integrator extracts the vital piece of information that the user needs from the downloaded Web pages and creates a database with the set of tables and respective attributes in the database. These tables are populated with the data extracted from the locally downloaded Web pages. This makes it easy for the user to query out his needs from the database.

The overall view of the system is as shown in Fig. 2.1. The system initializes a set of inputs for each block. These blocks are highlighted and framed according to the flow of data. The inputs that are needed for the start of the first block, i.e. the Web Data Extractor are fed in through the initialize phase. On receiving the inputs, the search engine performs its task of navigating to the page on the given URL and entering the criteria defined on the configuration file. This populates a page from which the actual download can start.

The process of extraction is to achieve the task of downloading the contents from the Web and to store them onto the local repository. Then onwards the process

Table 2.1 Basic notations

S	: Start Page URL	C, C_i	: Configuration File
l	: Depth of Recursion	W	: Set of Search Engine result Pages
$H(W)$	: Hyperlinks Set of W	CL	: Current Level
Lp	: Local Path to Hyperlinks	T_0	: Threshold for Similarity.

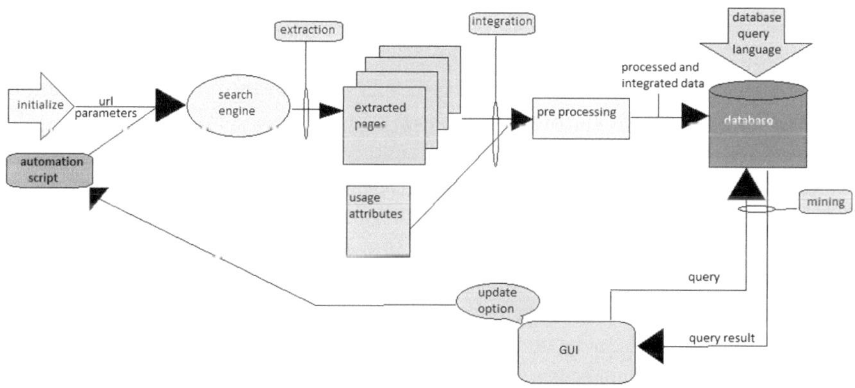

Fig. 2.1 Bluetooth SIG site

of extracting the pages iterates over, resulting in the outcome of offline Web pages (termed extracted pages). The extracted pages are available offline on a pre-described local repository. The pages in the repository can be browsed through in the same manner as that available on the Internet with the help of a Web browser. The noticeable thing here is that the pages are available offline, thereby are much faster to be accessed.

The outcome of the first block is to be chained with other attributes that are essential for the second block, namely the Database Integrator. The task of the second block is to extract the vital information from the extracted pages, process it and store it in accordance onto the database. The database is predefined in the set of attributes, together with the tables. This block needs the presence of the usage attributes that are extracted from the downloaded Web content. The usage attributes are mainly defined in a file termed as configuration file that the user needs to prepare before the execution. The file also contains the table to which the data extracted are to be added together with the table attributes and mapping.

On successful entries of the input, the integration block is able to accomplish the task of extracting the content from the Web pages. Since the Web pages do tend to remain in the same format as on the Internet, it is easy to be able to navigate across these pages as the links refer to the locally extracted pages. The extracted content is then processed to meet the desired data attributes that are listed on the database and the values are dumped onto it. This essentially creates rows of extracted information in the database. The outcome of second block results in tuples in the database that are essentially extracted out from the extracted pages got from the first block.

Now, that the database has been formed with the vital information contained in it, it would well be the task of the analyser, i.e. the third block referred to as GUI to provide the user with the functionality on how to deal with the contained data. The GUI provides for the interfaces that the user is able to achieve so as to obtain the data contained in the database, as and how he needs it. It acts as a miner that provides the user with the information obtained from the result of queries that are defined. The GUI also has options on referring the other blocks as requested by the user. This indeed interfaces all the three blocks, thereby providing the user with a better understanding and handling feature.

It is quite essential to know a few things that relate to the working of the entire system. First of all, it is very much essential that the inputs, initializations and the prerequisite data such as the usage attributes and/or configuration files that have been defined, do fall in a particular order and stick to the conventions defined on them. It is important to have the blocks to perform the tasks in order. The next thing is that, although the blocks do persist to function independently it is important that without the essential requirements, it is of no use to extract its functionality. Unless the prerequisites of the blocks are not met, the functionality that they accomplish is of no use. Similarly, to query out the needs of the user through the help of the GUI, it is essential that the data is available in the database. It is also essential that the first block and second block be run often and in unison so as to have an update on the current and ongoing dataset.

2.4 Mathematical Model and Algorithms

2.4.1 Web Data Extraction Using Similarity Function (WDES)

A connection has been established to the given url S and the page is processed with the parameters obtained from the configuration file C. On completion of this, we obtain the Web document that contains the links to all the desired contents that are obtained out of the search performed. The Web document contains individual sets of links that are displayed on each of the search results pages that are obtained. For example, if a search result obtained contains 150 records displayed as 10 records per page (in total 15 pages of information), we would have 15 sets of Web documents each containing 10 hyperlinks pointing to the required data. This forms the set of Web documents, W. That is,

$$W = \{w_i : 1 \leq i \leq n\}. \tag{2.1}$$

Each Web document $w_i \in W$ is read through to collect the hyperlinks that are contained in it, that are to be fetched to obtain the data values. We represent this hyperlink set as $H(W)$. Thus, we consider $H(W)$ as a whole set containing all the sets of hyperlinks on each page $w_i \in W$. That is,

$$H(W) = \{H(w_i) : 1 \leq i \leq n\}. \tag{2.2}$$

Then, considering each hyperlink $h_j \in H(w_i)$, we find the similarity between h_j and S, using Eq. 2.3.

$$SIM(h_j, S) = \frac{\sum_{i=1}^{min(nf(h_j),nf(S))} fsim(f_i h_j, f_i S)}{(nf(h_j) + nf(S))/2}. \quad (2.3)$$

where $nf(X)$ is the number of fields in X and $fsim(f_i h_j, f_i S)$ is defined as

$$fsim(f_i h_j, f_i S) = \begin{cases} 1\ if\ f_i h_j = f_i S \\ 0\ if\ f_i h_j \neq f_i S \end{cases} \quad (2.4)$$

The similarity $SIM(h_j,S)$ is the value that lies between 0 and 1, this value is used to compare with the defined threshold $T(0.25)$, we download the page corresponding to h_j to local repository if $SIM(h_j, S) \geq T$. The detailed algorithm of WDES is given in Algorithm 2.1.

Algorithm WDES navigates the search result page from the given $URL\ S$ and configuration file C and generates a set of Web documents W. Next, call the function $Hypcollection$ to collect hyperlinks of all pages in w_i, indexed by $H(w_i)$, page corresponding to $H(w_i)$ is stored in the local repository. The function $webextract$ is recursively called for each $H(w_i)$. Then, for each $h_i \in H(w_i)$, similarity between h_i and S is calculated using Eq. 2.3, if SIM(h_i, S) is greater than the threshold T_o, then page corresponding to h_i is stored and collect all the hyperlinks in h_i to X. Continue this process for X, until it reaches maximum depth l.

2.4.2 Web Data Integration Using Cosine Similarity (WDICS)

The objective of this algorithm is to extract data from the downloaded Web pages (those Web pages that are available in the local repository, i.e. output of WDES algorithm) into the database based on attributes and keywords from the configuration file C_i. We collect all the result pages W from local repository indexed by S, then $H(W)$ is obtained by collecting all hyperlinks from W, considering each hyperlink $h_j \in H(w_i)$ such that $k \in keywords$ in C_i. On existence of k in h_j, we populate the new record set $N[m, n]$ by passing page h_j and obtaining values defined with respect to the $attributes[n]$ in C_i. We then populate the old record set $O[m, n]$ by obtaining all values with respect to $attributes[n]$ in database.

For each record i, $1 \leq i \leq m$, we find the similarity between $N[i]$ and $O[i]$ using cosine similarity

$$SimRecord(N_i, O_i) = \frac{\sum_{j=1}^{n} N_{ij} O_{ij}}{\sqrt{\sum_{j=1}^{n} N_{ij}^2 \sum_{j=1}^{n} O_{ij}^2}}. \quad (2.5)$$

If similarity between records is equal to zero, then we compare each $attribute[j]$ $1 \leq j \leq n$ in the records and form $IntegratedData$ with use of $Union$ operation and store in the database.

$$IntegratedData = Union(N_{ij}, O_{ij}). \quad (2.6)$$

The detailed algorithm of WDICS is shown in Algorithm 2.2. The Algorithms WDES and WDICS, respectively, extract and integrate data in Depth First Search (DFS) manner. Hence, their complexity is O(n^2), where n is the number of hyperlinks in H(W).

Algorithm 2.1: Web Data Extraction using Similarity Function (WDES)

```
Input  : S: Starting Page URL, C: Parameter Configuration File, l: Level of Data
         Extraction, T_o: Threshold.
Output : Set of Webpages in Local Repository.
begin
    W=Navigate to Web document on Given S and automate page with C
    H(W)=Call: Hypcollection(W)
    for each H(w_i) ∈ H(w) do
        Save page H(w_i) on local Machine page P
        Call: Webextract(H(w_i), 0, pageppath)

Function Hypcollection(W)
begin
    for each w_i ∈ W do
        H(w_i)=Collect all hyperlinks in w_i
    return H(W)

Function Webextract(Z, cl, lp)
Input  : Z: set of URLs, cl: Current level, lp: local path to Z.
Output : Set of Webpages in Local Repository.
begin
    for each h_i ∈ Z do
        if SIM(h_i, S) > T_o then
            Save h_i to F_{h_i}
            X=collect URLs from h_i and change its path in lp
            if (cl < l) then
                Call: Webextract(X, cl + 1, pageppath of X)
```

2.5 Experiments

The algorithms are independent of usage and that they could be used on any given platform and dataset. The experiment was conducted on the Bluetooth SIG Website [18], The Bluetooth SIG (Special Interest Group) is a body that oversees the licensing and qualification of Bluetooth-enabled product. It maintains the database of all the qualified products in a database which can be accessed through queries on its website and hence is a domain-specific search engine. The qualification data contains a lot of parameters including company name, product name, any previous qualification

Algorithm 2.2: Web Data Integration using Cosine Similarity (WDICS)

```
Input  : S: Starting Page URL stored in local repository (output of WDES),
         C_i: Configuration File (Attributes and Keywords).
Output : Integrated Data in Local Repository.
begin
    H(w)=Call: Hypcollection(S)
    for each H(w_i) ∈ H(w) do do
        Call: Integrate(H(w_i))

Function Integrate(X)
Input  : X: set of URLs
Output : Integration of Values of Attributes Local Repository
begin
    for each h_i ∈ Z do
        if (h_i contain keyword) then
            new[m][n]=parse page to obtain values of defined attributes[n] in C_i
            old[m][n]=obtain all values of attribute[n] from repository
            for each record i do do
                if (SimRecord(new[i], old[i])==1) then
                    Skip
                else
                    for each attribute j do
                        if ( new[i][j] Not Equal to old[i][j] ) then
                            Integrateddata=union(new[i][j],old[i][j])
                    store Integarteddata in local repository
                X=collect all links for h_i
                if (X not equal to NULL) then
                    Call: Integrate(X)
```

used, etc. The main tasks for the Bluetooth SIG are to publish Bluetooth specifications, administer the qualification program, protect the Bluetooth trademarks and evangelize Bluetooth wireless technology. The key idea behind the approach is to collect and automate the collection of competitive and market information from the Bluetooth SIG site.

The site contains data in the form of a list that is displayed on the startup page. The display is formed based on the three types of listings; PRD 1.0, PRD 2.0 and EPL. The PRD 1.0 and PRD 2.0 are the qualified products and design list and EPL are the end product list being displayed. Each of the PRDs listed here are products that contain the specifications involved in them. The EPL are those that are formed in unison of the PRDs. Each PRD is identified by a QDID (an *id* for uniqueness) and each EPL is identified by the model. Each PRD may have many numbers of related EPLs and each EPL may have many numbers of related PRDs.

The listings as shown in Fig. 2.2, which contain links which navigate to the detailed content of the products that have been displayed here. The navigations on the page

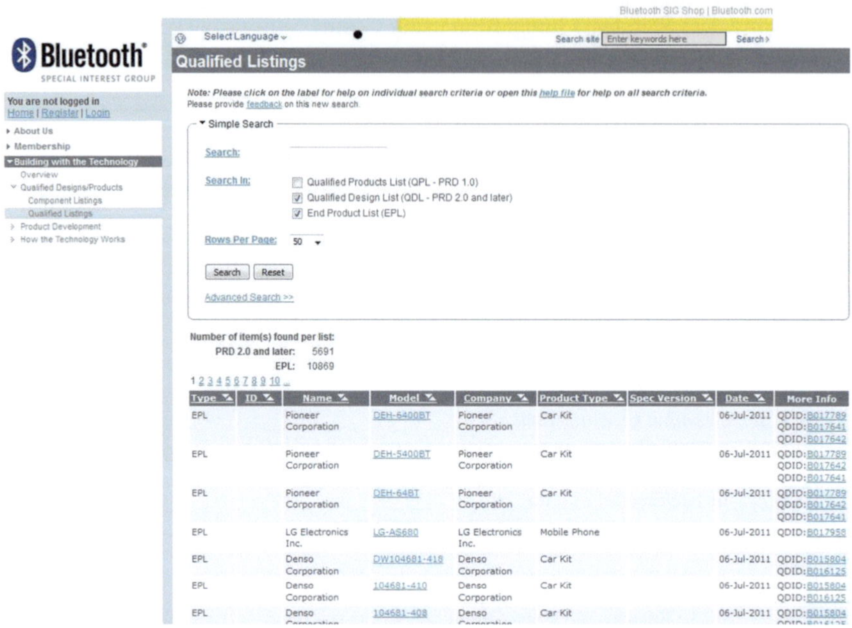

Fig. 2.2 SERP in bluetooth SIG site

reach to N number of pages before the case of termination. Thereby we may want to parse across the links to reach all the places to obtain the required data. Here, although we have all the data being displayed on the website, it is still not possible for us to prepare an analysis report based on the same. We want to play around with the data to get it to the form that we deserve it to be. Thereby, we incorporate the use of our algorithms to extract, integrate and mine the data from this site.

The experimental setup involves a Pentium Core 2 Duo machine with 2 GB RAM running windows. The algorithms have been implemented using a Java JDK and JRE 1.6, Java Parser with an access to an SQLite database and active broadband connection. Data has been collected from www.bluetooth.org, which lists the qualified products of Bluetooth devices. We have extracted 92 pages, with each page containing 200 records of information and data extraction was possible from each of these pages. Hence, we have a cumulative set of data for comparison based on the data extracted on the given attribute mentioned in the configuration file.

2.5.1 *Precision and Recall Versus Attributes*

Precision is defined as the ratio of correct pages and extracted pages and recall is defined as the ratio of extracted pages and the total number of pages. They are

Table 2.2 Performance evaluation between WDICS and DEPTA

Attributes	Total records (TR)	DEPTA		WDICS	
		Extracted records (ER)	Correct records (CR)	Extracted records (ER)	Correct records (CR)
Name	18234	18204	17325	18234	18234
Model	18234	17860	17010	18060	18060
Company	18234	18095	17208	18234	18198
Spec version	5508	5410	5016	5508	5426
Product type	18234	17834	17015	18234	18045
Total	78444	77403	73574	78270	77963
Recall=(ER/TR)*100		98.67%		99.77%	
Precision=(CR/ER)*100		95.05%		99.60%	

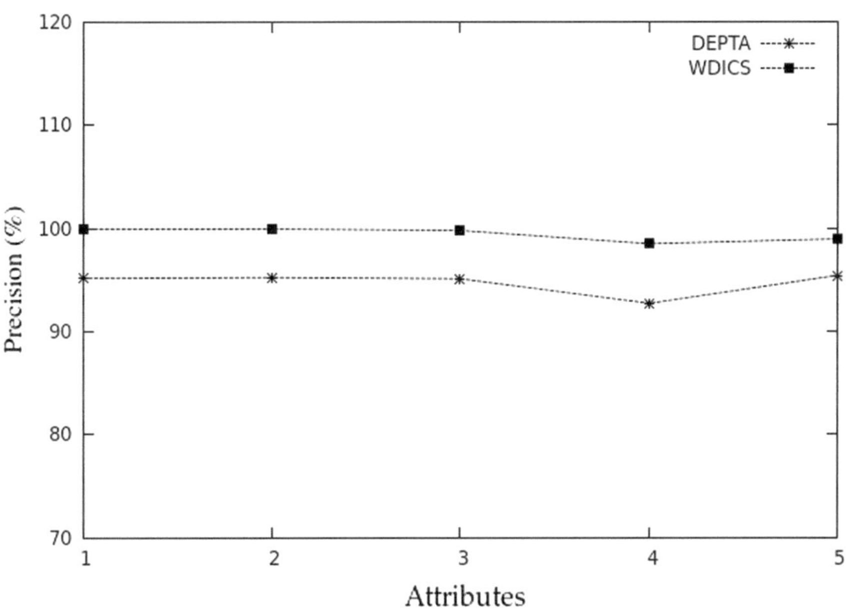

Fig. 2.3 Precision versus attributes

calculated based on the records extracted by our model, the records found by the search engine and the total available records on the Bluetooth website. For different attributes as shown in Table 2.2, the Recall and Precision are calculated and their comparisons are as shown in Figs. 2.3 and 2.4, respectively. It is observed that the Precision of WDICS increases by 4% and the Recall increases by 2% compared to DEPTA. Therefore, WDICS is more efficient than DEPTA because when an object is dissimilar to its neighbouring objects, DEPTA fails to identify all records correctly.

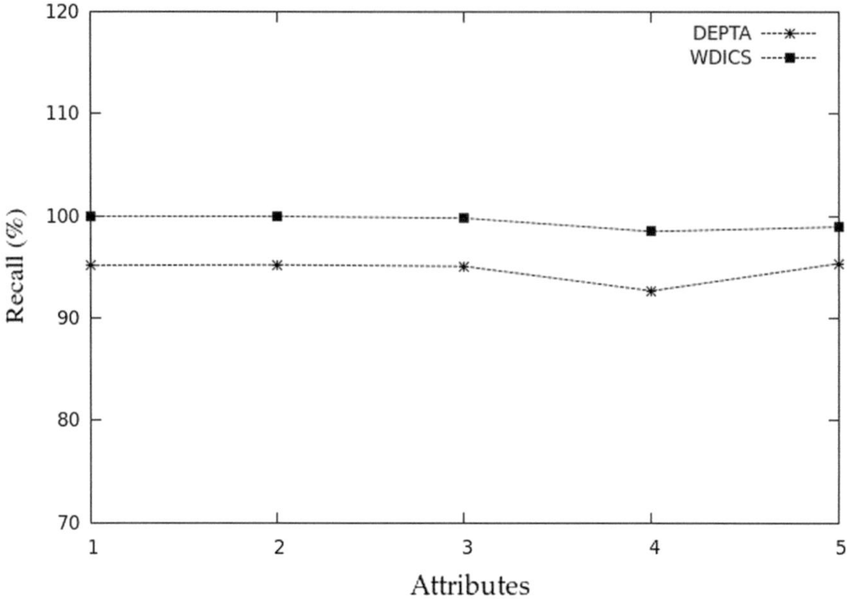

Fig. 2.4 Recall versus attributes

2.6 Summary

One of the major issues in Web Content Mining is the extraction of precise and meticulous information from the Web. In this chapter, we propose two similarity-based mechanisms; WDES to extract desired SERPs and store them in the local depository for offline browsing and WDICS to integrate the requested contents and enable the user to perform the intended analysis and extract the desired information. This results in faster data processing and effective offline browsing for saving time and resources. Our experimental results show that WDES and WDICS outperform DEPTA in terms of precision and recall. Further, different Web mining techniques such as classification and clustering can be associated with our approach to utilize the integrated results more efficiently.

References

1. Y. Zhai, B. Liu, Structured data extraction from the web based on partial tree alignment. J. IEEE TKDE **18**(12), 1614–1627 (2006)
2. G. Ananthanarayanan, S. Blagsvedt, K. Toyama, OWEB: a framework for offline web browsing, in *Fourth Latin America Web Congress, IEEE Computer Society* (2006), pp. 15–24
3. J. Myllymaki, Effective web data extraction with standards XML technologies, in *Proceedings of the 10th International Conference on World Wide Web* (2001), pp. 689–696

4. J. Yang, G. Shi, Y. Zheng, Q. Wang, Data extraction from deep web pages, in *IEEE International Conference on Computational Intelligence and Security* (2007), pp. 237–241
5. G.-S. Yin, G.-D. Guo, J.-J. Sun, A template-based method for theme information extraction from web pages, in *IEEE International Conference on Computer Application and System Modelling (ICCASM 2010)*, vol. 3 (2010), pp. 721–725
6. X. Liu, H. Li, D. Wu, J. Huang, W. Wang, L. Yu, Y. Wu, H. Xie, On web page extraction based on position of DIV, in *IEEE 4th ICCAE* (2010), pp. 144–147
7. N. Dalvi, P. Bohannon, F. Sha, Robust web extraction: an approach based on a probabilistic tree-edit model, in *Twenty-Eight ACM SIGMOD-SIGACT-SIGART Symposium on Principles of Database Systems* (2009), pp. 335–348
8. R. Novotny, P. Vojtas, D. Maruscak, Information extraction from web pages, in *IEEE/WIC/ACM International Conference on Web Intelligence and Intelligent Agent Technology Workshops* (2009), pp. 121–124
9. T. Nie, Z. Wang, Y. Kou, R. Zhang, Crawling result pages for data extraction based on URL classification, in *IEEE 7th Web Information Systems and Application Conference* (2010), pp. 79–84
10. N.K. Papadakis, D. Skoutas, K. Raftopoulos, STAVIES: a system for information extraction from unknown web data sources through automatic web wrapper generation using clustering techniques. J. IEEE TKDE **17**(12), 1638–1652 (2005)
11. C.-H. Chang, M. Ramzy, M.R. Girgis, A survey of web information extraction systems. J. IEEE TKDE **18**(10), 1411–1428 (2006)
12. N. Angkawattanawit, A. Rungsawang, Learnable crawling: an efficient approach to topic-specific web resource discovery, in *Proceedings of the 2nd International Symposium on Communications and Information Technology* (2002), pp. 97–114
13. K. Bharat, M.R. Henzinger, Improved algorithms for topic distillation in a hyperlinked environment, in *Proceedings of the 21st Annual International ACM SIGIR Conference on Research and Development in Information Retrieval* (1998), pp. 104–111
14. C. Aggarwal, F. Al-Garawi, P. Yu, Intelligent crawling on the world wide web with arbitrary predicates, in *Proceedings of the 10th International World Wide Web Conference* (2001), pp. 96–105
15. S. Chakrabarti, K. Punera, M. Subramanyam, Accelerated focused crawling through online relevance feedback, in *WWW* (ACM, 2002)
16. M. Ehrig, A. Maedche, Ontology-focused crawling of web documents, in *Proceedings of the 2003 ACM Symposium on Applied Computing* (2003), pp. 1174–1178
17. K.C. Srikantaiah, M. Suraj, K.R. Venugopal, S.S. Iyengar, L.M. Patnaik, Similarity based web data extraction and integration system for web content mining, in *Proceedings of the 3rd International Conference on Advances in Communication, Network and Computing Technologies, CNC 2012, LNICST* (2012), pp. 269–274
18. Bluetooth SIG Website, https://www.bluetooth.org/tpg/listings.cfm

Chapter 3
Mining and Cyclic Behaviour Analysis of Web Sequential Patterns

K. R. Venugopal and K. C. Srikantaiah

Abstract Understanding Web users' behaviour is an important criterion for improving the overall experience of Web users. Web Pattern Mining is one such field that helps us to mine useful behavioural patterns and draw conclusions from them after careful analysis. Efficient Web pattern mining is a challenge taking into consideration the enormous quantities of raw Web log data and explosive growth of information in the Web. In this chapter, we propose a novel algorithm called Bidirectional Growth based mining Cyclic Behaviour Analysis of Web sequential Patterns (BGCAP) that effectively combines these strategies to generate prefetching rules in the form of 2-sequence patterns with Periodicity and threshold of Cyclic Behaviour that can be utilized to effectively prefetch Web pages, thus reducing the users' perceived latency. In other words, BGCAP grow patterns bidirectionally along both ends of detected patterns and allows faster pattern growth with fewer levels of recursion thus eliminating unnecessary candidates and support for efficient pruning of invalid candidates. Due to these facts, BGCAP requires only (log n+1) levels of recursion for mining n Web Sequential Patterns. Our experimental results show that the Web Sequential Patterns and in turn prefetching rules generated using BGCAP is 5–10% faster for different data sizes and generates about 5–15% more prefetching rules than TD-Mine.

3.1 Introduction

Data Mining is the process of extracting useful information from a large repository of data. Web mining is one of the types of data mining and can be defined as the process of discovery and analysis of useful information from the data corresponding to World

K. R. Venugopal (✉)
Bangalore University, Jnana Bharathi, Bengaluru 560056, India
e-mail: venugopalkr@gmail.com

K. C. Srikantaiah
SJB Institute of Technology, BGS Health and Education City, Uttarahalli Main Road, Kengeri, Bengaluru 560060, India
e-mail: srikantaiahkc@gmail.com

K. R. Venugopal et al., *Web Recommendations Systems*,
https://doi.org/10.1007/978-981-15-2513-1_3

Wide Web. There are three major types of Web mining: (i) Web Structure Mining, (ii) Web Content Mining and (iii) Web Usage Mining. Web Structure Mining is the process of using graph theory to analyse a Web graph, where the nodes represent Web pages and the edges represent the hyperlinks among them. According to the type of Web structural data, Web structure mining can be divided into two categories: (i) extracting patterns from hyperlinks in the Web and (ii) mining the document structure, i.e. analysis of the DOM tree structure of the pages to describe HTML or XML tag usage. Web Content Mining is the process of mining actual content (text, image or multimedia) from the Web pages of the World Wide Web for information. Web content mining can be divided into two broad categories:

Agent-Based Approaches: Agent-based Web mining systems can further be divided into the following three categories: (i) Intelligent Search Agents: These Web agents are tools that interact with and learn the structure of unfamiliar Web pages and retrieve information from a variety of such sites using only general information about the domain. (ii) Information Filtering/Categorization Tools: A number of Web agents use various information retrieval techniques and characteristics of open hypertext Web documents to automatically retrieve, filter and categorize them. Criteria for categorization can be semantic information embedded in link structures and document content to create cluster hierarchies of hypertext documents, and structure an information space. (iii) Personalized Web Agents: This category of Web agents learn user preferences and discover Web information sources based on these preferences and those of other individuals with similar interests.

Database Approaches: Database approaches in Web content mining focus on techniques for organizing the semi-structured data on the Web into more structured collections of resources, and using standard database querying mechanisms and data mining techniques to analyse it. There are two techniques in this approach:

(i) Multilevel Database Systems: The main idea behind this technique is that the lowest level of the database contains semi-structured information stored in various Web repositories, such as hypertext documents. At the higher level, metadata or generalizations are extracted from lower levels and organized in structured collections, i.e. relational or page-oriented databases.

(ii) Web Query Systems: Web-based query systems utilize standard database query languages such as SQL for structural information about Web documents.

Web Usage Mining is the automatic discovery of user access patterns from Web logs stored in different servers. Servers collect large volumes of data in their daily operations, generated automatically and collected in server access logs. Other sources of user information include referrer logs which contain information about the referring pages for each page reference, and user registration or survey data gathered *via* scripts. Analysing such data can help in understanding the user's behaviour so that the server can improve its services like recommendation and Web personalization. Most Web analysis tools provide mechanisms for reporting user activity in the servers and various forms of data filtering. Using such tools, it is possible to determine the number of accesses to the server and to individual files, the times of visits and the domain names and URLs of users. These tools can be placed into two main categories such as follows:

Pattern Discovery Tools: These tools discover association rules and sequential patterns from server access logs. Sequential access patterns are essential for understanding and predicting the user's behaviour.

Pattern Analysis Tools: Once Web access patterns have been discovered, they need to be understood, visualized and interpreted so that the knowledge gained by the analysis can be utilized to improve the services offered by the Web servers.

The Internet is an extremely large collection of a network of networks, which in turn consist of Web servers which contain huge quantities of data, clients or end-user system which request information or services from the servers and finally client-side and server-side proxies which are additional systems that help and provide better communication amongst clients and servers. Such factors give rise to the necessity of creating intelligent systems that can effectively mine and analyse patterns and Web Usage Mining is one such technique. Mining from the Web includes integrating various data sources such as server access logs, referrer logs, user registration or profile information; and the importance of identifying user sessions or transactions from usage data, site topologies and models of user behaviour. Analysing such mined information results in predicting the users' behaviour and this knowledge in the form of prefetching rules is also extremely helpful in reducing users perceived latency and improving the quality of Web services.

WAP-Mine is one of the FP-growth-based algorithms for mining frequent Web access patterns from Web access database. But in the process of mining frequent Web access patterns, WAP-Mine produces many intermediate data which brings down the efficiency especially at lower support. TD-Mine, which is an extension of WAP-Mine, overcomes this problem and saves more space by reducing the amount of intermediate data generated. But, TD-Mine needs n levels of recursions to mine a pattern of length $(n + 1)$.

In our approach, we first perform preprocessing upon the raw Web logs to generate the session database of each Web user, which lists the Web users (IPs) along with their corresponding sessions. Then, we utilize a bidirectional pattern growth algorithm called UpDown Directed Acyclic Graph (UDDAG) [1] to generate Sequential Web access patterns from this session database and analyse them using Cyclic Model Analysis [2] to find out the Periodicity and Cyclic Behaviour of the mined 2-sequence patterns. The cyclic behaviour analysis can be used to generate Web prefetching rules. Constrains on Sequential Pattern Mining like date and time are specified, so that it returns interesting, more desirable, useful navigational patterns instead of huge and unwanted patterns.

In this chapter, we propose Bidirectional Growth based mining Cyclic behaviour Analysis of Web sequential Patterns (BGCAP) algorithm that utilizes UDDAG based on bidirectional pattern growth approach, which in turn focuses the search on a restricted portion of initial database to avoid expensive candidate generation and test step and is better on mining longer patterns. The strategies used in UDDAG are partitioning, projection and detection. UDDAG only needs *(log n+1)* levels of recursion thereby significantly reducing the execution time compared to TD-Mine. It takes minimal processing power for mining the complete set of sequential patterns in a large sequence database. These sequential patterns are analysed using Cyclic Model Analysis that deals with the tendency of certain sequential patterns to repeat themselves periodically after definite time intervals. This concept is greatly helpful in predicting future browsing patterns and prefetching Web pages aimed at certain user groups.

3.2 Related Works

Several techniques have been proposed for sequential pattern mining. They are mainly of two types: (i) A priori based (ii) Frequent Pattern growth (FP-growth) based. A priori based mining techniques such as a priori-all, GSP [3], SPADE [4], LAPIN-SPAM [5], LAPIN [6], scan the database multiple times. A n size pattern requires n scans of the database and hence these mining techniques are generally inefficient. FP-growth based mining techniques such as FreeSpan, BIDE [7], COBRA [8], PrefixSpan [9], UDDAG [1], etc., utilize a tree-based representation that reflects the original database and two scans are required to construct the tree. From this tree, the sequential patterns are derived without reference to the original database. Changes in the original database can easily be reflected in the tree by incremental analysis. These sequential pattern mining algorithms are used for mining Web access patterns from Web logs.

Cheng et al. [10] proposed an approach that combines a priori-all and clustering for sequential pattern mining to identify the user pattern and cluster users path patterns and make the similar users' cluster as one group in order to reduce the pattern that is effective for users in personalized service. But, it should be taken into account that not all patterns in a cluster may prove to be useful as they can produce erroneous conclusions after pattern analysis.

Gaol [11] explored habits of users using a priori-all algorithm, which first stores the original Web access sequence database for storing non-sequential data. This is based on the fact that the greater the number of combinations produced, the less likely the number of users who perform a combination of these and vice versa. While such an approach is simple and straightforward, such brute force tactics are obsolete as a priori-all algorithms are found to be least efficient with respect to sequential pattern mining.

Pei et al. [12] proposed Web Access Pattern tree (WAP-tree) for efficient mining of access patterns from Web logs. The Web access pattern tree stores highly compressed, critical information for sequential pattern mining. The WAP-tree registers all access sequence counts. There is no need for mining the original database any more as the mining process for all Web access patterns needs to work on the WAP-tree only. Therefore, WAP-mine needs to scan the access sequence database only twice. The height of the WAP-tree is one plus the maximum length of the frequent subsequences in the database. The width of the WAP-tree, i.e. the number of leaves of the tree, is the number of access sequences in the database. The size of the WAP-tree is much smaller than the size of access sequence database. It is shown that WAP-mine outperforms and has better scalability than GSP.

Xiaoqiu et al. [13] proposed the Improved WAP-tree in the form of highly compressed access sequences and introducing a subtree structure to avoid generation of conditional WAP-tree repeatedly and to generate maximal sequences. Improved WAP-tree excels traditional WAP-tree in time and space and shows better stability as the lengths of patterns vary. Mining frequent access sequences based on WAP-tree needs to scan transaction database only twice.

Yang et al. [14] designed an efficient algorithm Top-Down Mine (TD-mine) which makes use of the WAP-tree data structure for Web access pattern mining. WAP-tree can be traversed both top-down and bottom-up for the extraction of frequent access patterns. In TD-mine, a header table is used to traverse the tree from the root to the leaf nodes and mine patterns where the nodes are frequently accessed.

Liu et al. [15] proposed the Breadth-First Linked WAP-tree (BFWAP-tree) to mine frequent sequences which reflects parent–child relationship of nodes. The proposed algorithm builds the frequent header node links of the original WAP-tree in a Breadth-First fashion and uses the layer code of each node to identify the parent–child relationships between nodes of the tree. It then finds each frequent sequential pattern through progressive Breadth-First sequence search, starting with its first Breadth-First subsequence event. BFWAP avoids re-constructing WAP-tree recursively and shows a significant performance gain.

Vijayalakshmi et al. [16] designed an extended version of PrefixSpan called EXT-Prefixspan algorithm to extract the constraint-based multidimensional frequent sequential patterns in Web usage mining by filtering the dataset in the presence of various pattern constraints. EXT-PrefixSpan then mines the complete set of patterns but greatly reduces the efforts of candidate subsequence generation. This substantially reduces the size of the projected database and leads to efficient processing. EXT-PrefixSpan can be used to mine frequent sequential patterns of multidimensional nature from any Web server log file in the light of obtaining the frequent Web access patterns. However, EXT-PrefixSpan does not specify any particular constraint for consideration, i.e. it is highly generic.

Wu et al. [17] proposed the CIC-PrefixSpan, a modified version of PrefixSpan that mines and generates Maximal Sequential patterns by combining PrefixSpan and pseudo-projection. First, preprocessing is done to categorize the user sessions into human user sessions, crawler sessions and resource-download user sessions for efficient Web sequential pattern mining by filtering out the non-human user sessions, leaving the human user sessions and finding the transactions using Maximum Forward Path (MFP). By utilizing CIC-PrefixSpan, the memory space is reduced and generating duplicate projections to find the most frequent path in the users access path tree is also avoided. It is shown that CIC-PrefixSpan yields accurate patterns with high efficiency and low execution time compared to GSP and PrefixSpan. However, the frequent substructures within a pattern cannot be mined by CIC-PrefixSpan.

Verma et al. [18] designed a new pattern mining algorithm called Single-Level Algorithm for extracting behaviour patterns. These patterns are used to generate recommendations at run time for Web users. Single-Level Algorithm is designed keeping in mind the dynamic adaptation of focused websites that have a large number of Web pages. It combines preprocessing, mining and analysis to eventually predict the users' behaviour and hence is useful for specific websites and is highly scalable. It is shown to be more efficient than a priori algorithm. However, performing preprocessing on a very large Web log database can be time consuming and too cumbersome to be integrated with mining and analysis.

Nasraoui et al. [19] presented a framework for discovering and tracking evolving user profiles in real-time environment using Web usage mining and Web ontology. Preprocessing is first performed on the Web log data to identify user sessions. Then, profiles are constructed for each user and enriched with other domain-specific information facets that give a panoramic view of the discovered mass usage modes. This framework summarizes a group of users with similar access activities and consists of their viewed pages, search engine queries and inquiring and inquired companies. By mapping some new sessions to persistent profiles and updating these profiles, most sessions are eliminated from further analysis and focusing the mining on truly new sessions. However, this framework is not scalable.

Pitman et al. [20] modified the Bi-Directional Extension (BIDE) algorithm for mining closed sequential patterns in order to identify domain-specific rule sets for recommendation of pages and personalization for Web users in E-commerce. Individual supports are specified for each customer so that products can be recommended for individuals. Also, BIDE creates multidimensional sequences and further increases prediction for customers who do not explicitly specify their needs by using search functionality. However, additional strategies must be explored for identifying the most relevant sequential patterns without an exhaustive exploration of the search space bounded only by minimum support.

Masseglia et al. [21] proposed a Heuristic-based Distributed Miner (HDM), a method that allows finding frequent behavioural patterns in real time irrespective of the number of Web users. Navigational schemas, that are completely adaptable to the changing Web log data, are provided by HDM for efficient frequent sequence pattern mining. Based on a distributed heuristic, these schemas provide solutions for problems such as (i) discovering interesting zones (a great number of frequent patterns concentrated over a period of time) (ii) discovering super-frequent patterns and (iii) discovering very long sequential patterns and interactive data mining. However, the quality of the schemas can further be improved by adapting the candidate population.

Zhou et al. [22] designed an intelligent Web recommender system known as Sequential Web Access based Recommender System (SWARS) for sequential access pattern mining. Conditional Sequence mining (CS-mine) algorithm is used to identify frequent sequential Web access patterns. The access patterns are then stored in a compact tree structure, called Pattern tree, which is then used for matching and generating Web links for recommendations. SWARS has shown to achieve good performance with high satisfaction and applicability.

Yen et al. [23] address the issue of re-discovery of dynamic Web logs due to the obsolete Web logs as a result of deletion of users log data and insertion of new logs.

Incremental mining utilizes previous mining results and finds new patterns from the updated (inserted or deleted) part of the Web logs. A new incremental mining strategy called Incremental Mining of Web Traversal Patterns (IncWTP) is proposed, which makes use of an incremental updating algorithm to maintain the discovered path traversal patterns when entries are inserted or deleted in the database. This is achieved by making use of an extended lattice structure, which is used to store the previous mining results. However, the changes made to the website structure is not reflected in the lattice structure.

Zhang et al. [24] applied the Galois lattice to mine Web sequential access patterns by representing the paths traversed using graphs and compares the performance with that of a priori. Since the a priori-like algorithms frequently scan the entire transaction database to generate candidate patterns, Galois lattice reduces time complexity of closed sequential pattern mining as it needs only one scan.

Jain et al. [25] proposed a technique that employs Doubly Linked Tree to mine Web Sequential patterns. The Web access data available is constructed in the form of doubly linked tree. This tree keeps the critical mining related information in compressed format based on the frequent event count. It is shown that for low support threshold and for large database Doubly Linked Tree mining performance is better than conventional schemes such as a priori-all and GSP. However, Doubly Linked Tree does not work well in a distributed environment.

Jha et al. [26] proposed a Frequent Sequential Traversal Pattern Mining based on dynamic Weights constraint of Web access sequences (FSTPMW) to find the information gain of sequential patterns in session databases. The weight constraints are added into the sequential traversal pattern to control the number of sequential patterns that can be generated in addition to the minimum threshold. FSTPMW is efficient and scalable in mining sequential traversal patterns. However, it should be noted that FSTPMW does not consider levels of support along with the weights of sequential traversal patterns.

Wang et al. [27] proposed a Web personalization system that uses sequential access pattern mining based on CS-mine algorithm. The access patterns are stored in a compact tree structure called Pattern tree, which is then used for matching and generating Web links for recommendations. Pattern tree has shown to achieve good performance with accurate predictability.

Saxena et al. [28] integrated mining and analysis by proposing the One Pass Frequent Episode discovery (FED) algorithm. In this approach, significant intervals for each website are computed first and these intervals are used for detecting frequent patterns (Episodes). Analysis is then performed to find frequent patterns which can be used to forecast the users' behaviour. The FED algorithm is very efficient as it finds out patterns within one cycle of execution itself.

Oikonomopoulou et al. [29] proposed a prediction schema based on Markov Model that extracts sequential patterns from Web logs using website topology. Full coverage is achieved by the schema while maintaining accuracy of the prediction. Since Markov Models are infamous for their precision, the proposed prediction schema fails to deploy a more complex categorization method for each sequential pattern.

Rajimol et al. [30] proposed First Occurrence List Maximum (FOLMax-mine) for mining maximal Web access patterns based on FOL-Mine. It is a top-down method that uses the concept of first occurrence to reduce search space and improve the performance. This is achieved by finding out the Maximal Frequent Path in the patterns generated from the Web logs.

3.3 System Architecture

3.3.1 Problem Definition

Given a Web log database W and the set of pages $P = \{p_i\colon 1 \leq i \leq n\}$ in a Web server, a *session* is a subset of P, denoted by $(p_1, p_2, \ldots p_k)$, where $p_i \in P$, $i \in \{1, \ldots, k\}$. Here, the parentheses are omitted for a session with one page only. A *Web access sequence* q is a list of sessions, denoted by $< q_1 q_2 \ldots q_m >$, where q_i is a session and $q_i \subseteq P$, $i \in \{1, \ldots, m\}$. The number of sessions in q is called the length of q.

Given two access sequences $x =< x_1 x_2 \ldots x_j >$ and $y =< y_1 y_2 \ldots y_k >$, x is said to be a *subsequence* of y and y a *supersequence* of x if $k \geq j$ and there exists integers $1 \leq i_1 < i_2 < \cdots < i_j \leq k$, such that $x_1 \subseteq y_{i1}, x_2 \subseteq y_{i2}, \ldots, x_j \subseteq y_{ij}$. Here, x is also contained in y which is denoted by $x \alpha y$. A *session database* is a set of tuples $< ip, q >$, where ip is the users IP address which is used as a sequence identifier and q is the users access sequence. A tuple $< ip, q >$ is said to contain an access sequence α if αq.

The support of a subsequence α, denoted by Support (α), is the number of sequences for which α is a subsequence. A subsequence α is said to be a Web sequential access pattern, when its support is greater than user-specified minimum support ($MinSup$), i.e. $Support(\alpha) \geq MinSup$. Given a session database and MinSup, the objective is to extract Web sequential access patterns using Bidirectional Pattern Growth algorithm, analyse them using Cyclic Model Analysis to determine the periodicity and cyclic behaviour of the 2-sequence patterns and employ them as prefetching rules. *Assumptions*: During the preprocessing operation of the Web logs, it is assumed to be sufficient to consider only those URLs whose domains contain text and images only. Web sequential patterns pertaining to multimedia Web pages need not be mined and hence are filtered out during preprocessing of the raw Web logs.

Basic Definitions

Web Prefetching: Web Prefetching is the process of fetching Web pages from the Web server before they are actually requested by the users. Periodicity [31]: The periodicity P of the 2-sequence $< p_i p_j >$ is defined as the time period t after which p_j shall be accessed periodically after p_i has been accessed. Example: The periodicity

P between a pair of Web pages (p_i, p_j) is 10 ms indicates that the page p_j is accessed periodically with period 10 ms after page p_i has been accessed.

Tendency [31]: The tendency between a pair of Web pages (p_i, p_j) is defined as the line in the trend graph that determines whether the cyclic behaviour of the 2-sequence $< p_i p_j >$ is increasing or decreasing.

Cyclic Behaviour [31]: The cyclic behaviour C of the 2-sequence $< p_i p_j >$ is defined as the time that gives the stopping criteria for accessing page p_j after p_i has been accessed. It is calculated using Periodicity and Tendency. The periodicity is increased by adding the value to itself each time the page is accessed. The page will not be accessed after periodicity has reached the limit of cyclic behaviour, i.e. $P \leq C$. Example: The periodicity P between a pair of Web pages (p_i, p_j) is 10 ms, the trend line is decreasing and the cyclic behaviour $C = 50$ ms indicates that accessing of page p_j repeats for every 10 ms after accessing the page p_i. This ends when the time t reaches 50 ms.

The system architecture consists of the following components, (i) Web logs, (ii) Preprocessing Engine, (iii) Encoder, (iv) Sequential Pattern Miner, (v) Pattern Analyser and (vi) Prefetching Rules is as shown in Fig. 3.1.

Web Logs: A Web log is a large database stored in Web servers that contain details of the transactions of Web users. There are many fields in the Web log database, which conform to either Common Log Format (CLF) or the Extended Common Log Format (ECLF) as shown in Figs. 3.2 and 3.3, respectively. The fields specified by this format are IP address of the destination page, destination URL, code which denote the packet number, protocol which specify the type of network protocol used (e.g. TCP), method which specifies the type of HTTP method used (e.g. GET, POST),

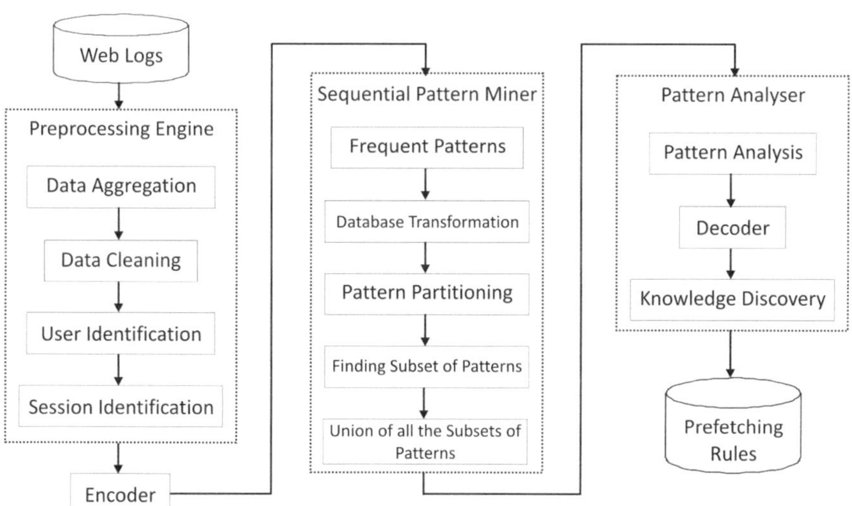

Fig. 3.1 System architecture

<IP_addr><base_url><date>
<method><file><protocol>
<code><bytes>

Fig. 3.2 Common log format (CLF)

<IP_addr><base_url><date>
<method><file><protocol>
<code><bytes><referrer>

Fig. 3.3 Extended common log format (ECLF)

Fig. 3.4 Access log from the Web server

Web-proxy,debug,packet1307775248.816 363 30.0.1.2 TCP_MISS/200 960 GET http://www.facebook.com/ajax/typehead/search.php?-DIRECT/66.220.146.32 application/x-javascript in 11-Jun 12:25;6.76 from 30.0.7.254

web-proxy, debug, packet 1307775249.609 586 30.0.0.223 TCP_MISS/200 397 POST http://channeltvunetworks.com/all-DIRECT/ 38.103.62.170 text/html in 11-Jun 12:25:7.69 from 30.0.7.254

type which specifies the file type, date which gives the date and time when the page was accessed, referrer which gives the IP address of the client which requested the page, and finally the size of the page in bytes. An extract from a Web log is as shown in Fig. 3.4. Web logs are very useful for predicting the behaviour of the clients and hence different mining techniques can be used to find and extract interesting browsing patterns of the users.

Preprocessing Engine: This component aggregates the Web logs data from different sources and produces session database as the result using the following steps:

Data Aggregation: The Web logs from different sources with different formats are extracted and integrated into a single database so that they pertain to a single format with no redundancy.

Data Cleaning: In this step the aggregated data is cleansed, i.e. useless records such as URLs containing images, multimedia, scripts and entries corresponding to crawlers are removed and only human-initiated entries (i.e. URLs ending with HTM, HTML, XHTML, PHP and JSP) are retained. Only a few of these fields are important for the mining process and hence after extracting or collecting only the important fields such as the User IP Addresses (referrers), URLs and date and the type of file in the URL (whether text, image or script), the rest can be ignored.

User Identification: The identity of a user is not a prerequisite for Web usage mining. However, it is necessary to differentiate users to identify sessions. The cleaned database is grouped by different IP addresses and sorted by Date and Time for each IP for identifying different users. As the database is used to store this information, simple queries can be used to achieve this operation. By doing so, we can identify individual users and also find out the total number of users.

Session Identification: Session Identification is the process of partitioning the user activity records of each user into sessions in order to reconstruct the actual sequence of actions done by each user. A session is a package of activities that consists of a user's navigation history. These sessions are then aggregated to create a session database. The user activity records are divided into sessions by assigning unique identifiers for each session. Each user is assigned a separate session and also a separate session is assigned for the same user if the user exceeds a certain threshold of time (e.g. 15 min).

Encoder: Encoding is the process where each URL in the preprocessed database is assigned a unique identifier which is a non-negative integer. This is done as it would be too cumbersome to mention the URLs by their domain names in the session database. For this purpose, a table consisting of the list of URLs can be maintained where they are mapped to their respective identifiers. User Identification, Session Identification and Encoding of URLs are all shown in the example in Fig. 3.5. After encoding, the session database, which consists of users and their sessions, is generated and an example is shown in Fig. 3.6.

Pattern Miner: Pattern Mining is used to find hidden patterns from a large database when a minimum threshold of occurrence (MinSup) has been specified. The Pattern Miner utilizes UDDAG that grow patterns bidirectionally along both ends of detected patterns and allows faster pattern growth with fewer levels of recursion. UDDAG eliminates unnecessary candidates and supports efficient pruning of invalid candidates. This represents a promising approach for applications involving searching in large spaces like Web Sequential Pattern Mining.

Pattern Analyser: Pattern Analysis is the process to study and conduct the analysis of the results obtained from the Web sequential access patterns derived by the miner. The analyser utilizes CMA to find out the periodicity and cyclic behaviour of all the 2-sequence patterns mined. Decoding is the process where the identifiers are

IP	DATE_TIME	URL	URL_ID	SESSION_ID
30.0.7.231	12-Jun-11 12:36:18 AM	http://dnl-02.geo.kaspersky.com/diffs/t	253	2
30.0.7.231	12-Jun-11 12:54:54 AM	http://myinfo.any-request-allowed.com	480	3
30.0.7.231	12-Jun-11 12:56:51 AM	http://myinfo.any-request-allowed.com	480	3
30.0.7.231	13-Jun-11 12:16:08 AM	http://myinfo.any-request-allowed.com	480	4
30.0.7.231	13-Jun-11 12:37:37 AM	http://myinfo.any-request-allowed.com	480	5
30.0.0.62	11-Jun-11 12:41:54 AM	http://www.orkut.com/Glogin?	1071	6
30.0.0.62	11-Jun-11 12:41:54 AM	http://www.phoenixads.co.in/delivery/	1103	6
30.0.0.62	11-Jun-11 12:41:59 AM	http://channel.tvunetworks.com/list/all	185	6
30.0.0.62	11-Jun-11 12:48:45 AM	http://indiarailinfo.com/main/Getbanner	397	6
30.0.0.62	11-Jun-11 12:50:07 AM	http://clients1.google.co.in/generate_2	194	6
30.0.1.3	11-Jun-11 12:00:53 AM	http://metric.ind.rediff.com/www.rediff	465	7
30.0.1.3	11-Jun-11 12:00:55 AM	http://www.erail.in/parter/GetStation	909	7
30.0.1.3	11-Jun-11 12:00:57 AM	http://ad.doubleclick.net/adi/N6404.272	88	7

Fig. 3.5 Preprocessed data

IP	ACCESS_SEQ
30.0.5.108	< 1119 324 566 67 942 >,<187 187 98 194>,< 878 878 187 56 443 242 449 942 718 >
20.0.0.98	< 67 826 185 185 293 1059 1192 990 946 1218 189 1006 960 98 1219 316 324 926 1215 67 962 98>
30.0.0.114	< 597 939 653 >
30.0.0.142	< 324 641 649 185 983 185 722 53 806 185 951 189 324 950 951 740 676 720 1185 797 721 797 >
30.0.1.19	< 324 >
20.0.0.88	<739 739 274 272 277 279 850 654 561 434 627 324 194 658 1241 277 275 >
30.0.0.151	< 295 294 87 55 21 >,< 1074 1074 946 >
30.0.0.153	< 388 940 800 >
30.0.0.111	< 324 893 194 575 >
40.0.1.35	< 187 243 1004 185 185 919 388 185 397 71 395 1003 >
30.0.1.17	< 315 741 1242 >
30.0.4.189	< 1103 898 >
30.0.0.50	< 238 98 >
30.0.0.110	< 563 196 874 717 779 103 566 >

Fig. 3.6 Session database

replaced with their corresponding URLs. In the Knowledge Discovery process, the analysed data is transformed into Web prefetching rules and sent to the Prefetching Rule Depository so that they can be used to prefetch and cache Web pages and reduce the round-trip delay experienced by the users.

Prefetching Rule Depository: This component is a large database consisting of prefetching rules pertaining to the requested Web pages. After the pattern analysis, each 2-sequence pattern having periodicity and cyclic behaviour are stored here in the form of prefetching rules and are triggered when the first page in the sequence is accessed by a user.

3.4 BGCAP Algorithm

Given a raw Web log database W, we first perform the preprocessing and generate the session database SD as shown in Table 3.1. The session database consists of a set of tuples, with each tuple consisting of the user (IP) and the access sequence of that user. From Table 3.1, the user with IP address 1.0.1.2 has accessed the Web pages A, B, C and D with the sequence $< A(A, C)BD >$, where $\{A, B, C, D, E, F\}$is the set of unique Web pages accessed by different users. Here (A, C) in the above sequence denotes that in a single session, the two Web pages A and C were visited by the user in that order, otherwise the user is assumed to visit a single page per session if the parentheses () are not specified.

Web sequential patterns are mined from the session database SD using updated version of UDDAG [1]. This mining technique is basically a divide-and-conquer approach, which tries to construct the patterns simultaneously along both directions of a Directed Acyclic Graph (DAG). The approach consists of three main steps: (i) Database Transformation (ii) Pattern Partitioning and (iii) Finding subsets of Patterns.

Database Transformation is used to remove infrequent pages, i.e. those pages that do not have $MinSup = 2$. From Table 3.1, the frequent items with $MinSup \geq 2$ are

Table 3.1 Example sessions database

User (IP)	Sequences
1.0.1.2	$< A(A, C)B\ D >$
1.0.1.3	$< A\ D(E, F) >$
1.0.1.4	$< (B, D)C\ F >$
1.0.1.5	$< (C, E)(A, B, C, D) >$
1.0.1.6	$< A\ B\ C\ D\ E >$

Table 3.2 After database transformation

User (IP)	Sequences
p	$< 1\ 2\ 3\ 6 >$
q	$< 1\ 6(7, 8) >$
r	$< 4\ 5\ 8 >$
s	$< (5, 7)(1, 3, 5, 6) >$
t	$< 1\ 3\ 5\ 6\ 7 >$

(A), (B), (C), (D), (E), (F), (A, C), (B, D). Since we require only these patterns, the remaining can be eliminated by substituting these patterns with non-negative integers like $(A) - 1$, $(A, C) - 2$, $(B) - 3$, $(B, D) - 4$, $(C) - 5$, $(D) - 6$, $(E) - 7$, $(F) - 8$. For the simplicity of representation, we assign each IP a unique identifier as well. The transformed database is as shown in Table 3.2.

Next, the patterns have to be partitioned into projected databases for each pattern, denoted by ^{n}D, where n represents the frequent item. By partitioning, we select only that tuples in which n is present, called projected database for item n and ignore the rest. So, Table 3.2 is partitioned into m projected databases as shown in Table 3.3.

The third and final step, finding subsets of patterns, is not as straightforward as the previous steps. If WP is the set of all sequential patterns mined from W with $MinSup = 2$, then let WP_1, WP_2, ..., WP_8 be the subsets of patterns, i.e., the pattern mined from ^{1}D, ^{2}D, ..., ^{8}D respectively. If the condition $|^n D| \geq 2$ is not met then such projected databases can be ignored. Here, the projected databases ^{2}D and ^{4}D can be ignored as they contain only one tuple each and hence do not contribute frequent patterns. So we now have to find WP_1, WP_3, WP_5, WP_6, WP_7 and WP_8.

Let us consider ^{6}D. As seen in Table 3.3, the projected database for 6 contains four tuples with IPs p, q, s and t. We have to find out the sequential patterns in ^{6}D by recursively partitioning the projected database into prefix and suffix databases until no frequent patterns are found. In the first round of partitioning, we split the projected database into Prefix (Pre (^{6}D)) and Suffix (Suf (^{6}D)) subsets, each containing the tuples with the prefix and suffix sequences, respectively, pertaining to 6 as shown in Table 3.4.

Again, the frequent items, i.e. the patterns in Pre (^{6}D) are $< 1 >$, $< 3 >$ and $< 5 >$, and the pattern in Suf (^{6}D) is $< 7 >$. So, Pre (^{6}D) is further split into PP_1, PP_3 and PP_5, and Suf (^{6}D) is split as PS_7 and this process continues recursively

Table 3.3 Projected databases

^1D	^2D
p:< 1 2 3 6 >	p:< 1 2 3 6 >
q:< 1 6(7, 8) >	
s:< (5, 7)(1, 3, 5, 6) >	
t: < 1 3 5 6 7 >	
^3D	^4D
p: < 1 2 3 6 >	r: < 4 5 8 >
s: < (5, 7)(1, 3, 5, 6) >	
t: < 1 3 5 6 7 >	
^5D	^6D
r: < 4 5 8 >	p:< 1 2 3 6 >
s: < (5, 7)(1, 3, 5, 6) >	q:< 1 6(7, 8) >
t: < 1 3 5 6 7 >	s: < (5, 7)(1, 3, 5, 6) >
	t: < 1 3 5 6 7 >
^7D	^8D
q:< 1 6(7, 8) >	q:< 1 6(7, 8) >
s:< (5, 7)(1, 3, 5, 6) >	r: < 4 5 8 >
t: < 1 3 5 6 7 >	

Table 3.4 Prefixes and suffixes for ^{6}D

Sequence sets	Frequent patterns
Pre (^{6}D):{< 1 2 3 >, < 1 >, < (5, 7) >, < 1 3 5 >, }	< 1 >, < 3 >, < 5 >
Suf (^{6}D):{< (7, 8) >, < 7 > }	< 7 >

until no frequent patterns are available. The prefix and suffix databases pertaining to PP_1, PP_3, PP_5 and PS_7 are as shown in Table 3.5.

From Table 3.5, it can be seen that only two patterns < 3 > and < 1 > are frequent and no other frequent patterns exist in projected database for < 3 > and < 1 >. Hence, the partitioning stops and the DAG is constructed as shown in Fig. 3.7 to find all the patterns of ^{6}D.

In Fig. 3.7, < 6 > is the root node of the DAG and it contains occurrence set $\{p\ q\ s\ t\}$ to show that the pattern < 6 > occurs in the tuples with IPs p, q, s and t. In a DAG, the Up-children and Down-children denote the prefixes and suffixes of the root node, respectively. Here, the root node < 6 > has three Up-children and one Down-child. Pre (^{6}D) yielded < 1 >, < 3 > and < 5 > and hence these are prefixes of < 6 > and < 1 6 >, < 3 6 > and < 5 6 > are frequent patterns as they occur in tuples with IPs $\{p\ q\ t\}$, $\{p\ t\}$ and $\{s\ t\}$, respectively. Suf (^{6}D) yielded < 7 > and is a suffix of < 6 > and hence < 6 7 > is a frequent pattern as it occurs in tuples with IPs $\{q\ t\}$. From Suf (PP_1), suffix of < 1 > is < 3 > and from Pre (PP_3), the prefix

Table 3.5 Prefixes and suffixes of prefix and suffix databases of ^{6}D

Sequence sets	Frequent patterns
Pre (PP_1):$\{ <> \}$	–
Suf (PP_1):$\{< (2, 3) >, < (3, 5) > \}$	$< 3 >$
Pre (PP_3):$\{ < 1\,2 >, < 1 > \}$	$< 1 >$
Suf (PP_3):$\{ < 5 > \}$	–
Pre (PP_5):$\{ < 1\,3 > \}$	–
Suf (PP_5):$\{ <> \}$	–
Pre (PP_7):$\{ <> \}$	–
Suf (PP_7):$\{<> \}$	–

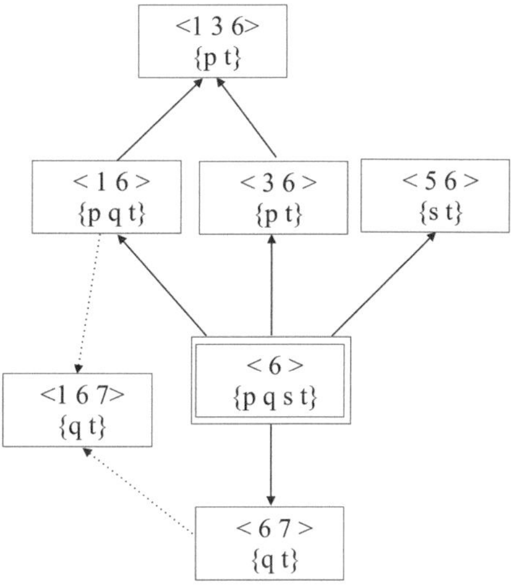

Fig. 3.7 The example DAG

of $< 3 >$ is $< 1 >$, therefore, $< 1\,3\,6 >$, the upmost child node, is a frequent pattern as it occurs in tuples with IPs $\{p\ t\}$. The patterns in an Up-child and Down-child of a DAG can be combined to form a new node only if the number of tuples of the occurrence sets of the Up-child and Down-child is greater than or equal to $MinSup$. So, the node $< 1\ 6\ 7 >$ has one Up-parent $< 1\ 6 >$ and one Down-parent $< 6\ 7 >$ and as it is a frequent pattern in $\{q\ t\}$, $< 167 >$ is a valid sequential pattern.

Therefore, the complete set of patterns in ^{6}D is, $WP_6 = \{< 6 >, < 1\ 6 >, < 3\ 6 >, < 5\ 6 >, < 6\ 7 >, < 1\ 3\ 6 >, < 1\ 6\ 7 >\}$. Similarly, $WP_1 = \{< 1 >, < 1\ 3 >, < 1\ 6 >, < 1\ 3\ 6 >, < 1\ 6\ 7 >\}$, $WP_3 = \{< 3 >, < 1\ 3 >, < 3\ 6 >, < 1\,3\,6 >\}$, $WP_5 = \{< 5 >, < 5\,6 >\}$, $WP_7 = \{< 7 >, < 1\,7 >, < 6\,7 >, < 1\,6\,7 >\}$ and $WP_8 = \{< 8 >\}$. The complete set of patterns in the session database is $WP = \{< 1 >, < 3 >, < 5 >, < 6 >, < 7 >, < 8 >, < 1\,3 >, < 1\,6 >, < 1\,7 >, < 3\ 6 >, < 5\ 6 >, < 6\ 7 >, < 1\ 3\ 6 >, < 1\ 6\ 7 >\}$. Maximal forward references save memory and can be used to generate all the above sequential patterns for WP

and so WP (max) is $\{<8>, <5\ 6>, <1\ 36>, <1\ 6\ 7>\}$. As this algorithm focuses on deriving prefetching rules, we use WP (max) to generate all 2-sequence Web access patterns. Set of all 2-sequence patterns are $\{<1\ 3>, <1\ 6>, <3\ 6>, <1\ 7>, <6\ 7>, <5\ 6>\}$. These 2-sequence patterns are analysed using Cyclic Model Analysis (CMA) [2] to find out the periodicity and cyclic behaviour of these sequences in the sequence database D. After analysis, the set of prefetching rules PR are derived from the periodicity and cyclic behaviour and stored in the server.

Algorithm 3.1: BGCAP Algorithm

```
Purpose: To find Periodicity and Tendency of sequential patterns.
Input  : W (Web Log Database)
Output : PR (Set of Prefetching Rules)
begin
    F = Cleaning(W)
    SD = SessionIdentifier(F,THRESHOLD)
    WP = Bidirection Pattern GrowthP(SD,minsup)
    PR = Pattern Analysis(WP)
Cleaning(W)
begin
    for each l ∈ W do do
        if (URL in l contains (js, css)) then
            ignore
            else
                insert l into F
SessionIdentifier(F,THRESHOLD)
Purpose: To identify Sessions
Input: F- Cleaned database and sorted Web log entries according to IP address
        and time, Threshold time limit for each session (L_i.ip IP
         address at record L_i and L_i.t Date/Time entry at record L_i )
Output : F- with Sessions
begin
    Vsession_id = 0
    for each l ∈ f do do
        if (l_i.ip != l_{i+1}.ip) then
            session_id++
            l_i.sid = session_id
            else if (l_i.ip == l_{i+1}.ip and (l_{i+1}.t - l_i.t) > threshold) then
                session_id++
                l_i.sid = session_id
            else
                l_i.sid = session_id
```

For example, if the 2-sequence pattern $<3\ 6>$ occurs frequently in D with a Periodicity of 10 s, that means the user accesses $<6>$ 10 s after he has accessed $<3>$. If this behaviour repeats itself and then stops after 80 s, then it is said to be the threshold of Cyclic Behaviour of $< 3\ 6 >$ after which this pattern will not repeat again. Using this knowledge, the Web page $< 6 >$ can be prefetched for the user before the 10th s and stored in the cache for future references and hence reduce his perceived latency. The Web page can be deleted from the cache when the Cyclic Behaviour of 80 s has been reached. The algorithm summarizes the above processes and is shown in Algorithm 3.1.

3.5 Experiments

The algorithm BGCAP has been implemented using Java language using NetBeans 6.9.1 platform on MSNBC dataset in a Pentium Dual-Core processor environment, with a 2 GB Memory and 100 GB HDD. The MSNBC Web log data comes from Internet Information Server (IIS) logs for msnbc.com and news-related portions of *msn.com* for 989818 users. Each sequence in the dataset corresponds to page views of a user during that twenty-four hour period. Each event in the sequence corresponds to a user's request for a page. The page requests served *via* caching mechanism are not recorded in the server logs and hence, not present in the data.

Experiments are pursued to compare the efficiency of BGCAP and TD-Mine. BGCAP demonstrated satisfactory scale-up properties with respect to various parameters such as the total number of Web access sequences, the total number of pages, the average lengths of sequences. The following comparisons show that BGCAP outperforms TD-Mine in quite a significant margin and has better scalability than TD-Mine.

3.5.1 *Data Size Versus Run Time*

The data size is the number of transactions of the input session database and it is a significant factor that affects the performance of BGCAP. This is demonstrated in Fig. 3.8, where we show how different data sizes have different execution times. The higher the number of transactions, the more time it would take to process the data and generate patterns. However, as seen in Fig. 3.8, TD-Mines execution time is moderately higher than that of BGCAP. This is because of the Bidirectional pattern growth approach adopted by BGCAP that reduces the run time by 5–10% than that of TD-Mine. However, as the data size increases to the order of about a million transactions, the run time of both the approaches will tend to be the same since BGCAP must recursively generate prefixes and suffixes for all the frequent items mined.

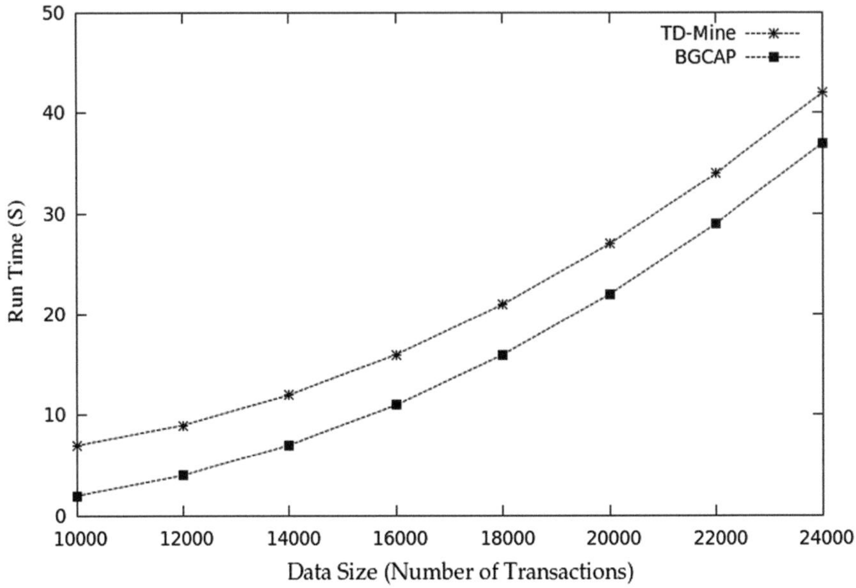

Fig. 3.8 Data size versus run time

3.5.2 Threshold Versus Run Time

The minimum support threshold ($minSup$) is another important factor that affects the performance of BGCAP. This is demonstrated in Fig. 3.9 where we show how different support thresholds (for a fixed 2000 transactions) have different execution times. As the data size is fixed, the mining only depends on the $minSup$ values. As the $minSup$ value increases, the execution time gradually decreases as it would be a smaller number of patterns that are to be mined. In Fig. 3.9, the graph shows that TD-Mine takes more time to generate patterns compared to BGCAP and they both tend towards the same run time as $minSup$ increases. It is observed that the approach adopted by BGCAP reduces the run time by 10–15% than that of TD-Mine.

3.5.3 Threshold Versus Number of Patterns

Next, we mine Web sequential patterns for different thresholds for a fixed data size of 2000 transactions. Figure 3.10 shows how many patterns can be mined for different support thresholds using BGCAP and TD-Mine. We can see that the number of patterns mined using BGCAP is significantly higher (5–15%) than that of TD-Mine as shown in Fig. 3.10. This is because of the Bidirectional pattern growth approach adopted by BGCAP that is more scalable and accurate than TD-Mine. However, as

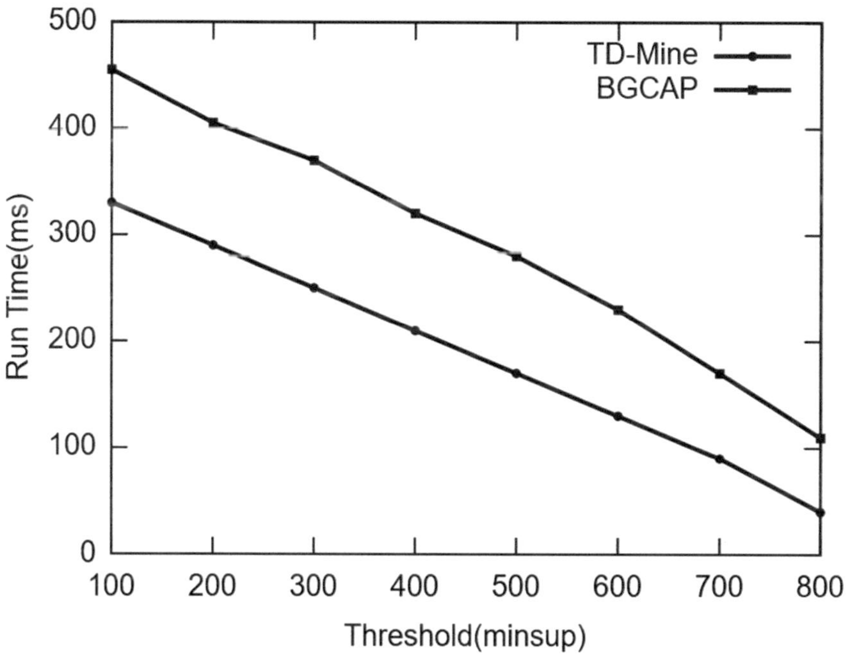

Fig. 3.9 Threshold versus run time

the minSup increases to the order of 1000 frequent items, the number of patterns generated using both these approaches will tend to be the same as such patterns rarely occur for such large values of threshold.

After generating the Web 2-sequence patterns, BGCAP analyses them using CMA to derive the prefetching rules in terms of periodicity and cyclic behaviour. Consider the same example in Sect. 3.4, where we derived the 2-sequence patterns: $\{<1\,3>, <1\,6>, <3\,6>, <1\,7>, <6\,7>, <5\,6>\}$. After analysis, we get the following information as shown in Table 3.6.

From Table 3.6, for the 2-sequence pattern $<1\,3>$, the Periodicity is 9 s which means Web page (3) will be accessed by a user 9 s after Web page (1) has been accessed and so it can be prefetched from the server and stored in the client system before that users actually requests the page (3). As the prefetched Web page has already travelled from the server, the users perceived delay is insignificant as the page (3) is simply fetched from the client systems memory itself when the user requests the prefetched Web page. The threshold of cyclic behaviour for $<1\,3>$ is 57 s which means that this behaviour stops after 57 s, i.e. the user does not request page (3) after page (1) has been accessed, therefore implying that (3) need not be prefetched after the cyclic behaviour has been reached and hence it can be removed from the clients memory. Thus, it reduces network traffic and saves resources by not prefetching Web pages that shall never be requested by the users. Similarly, the prefetching rules can be generated for n-sequence Web patterns.

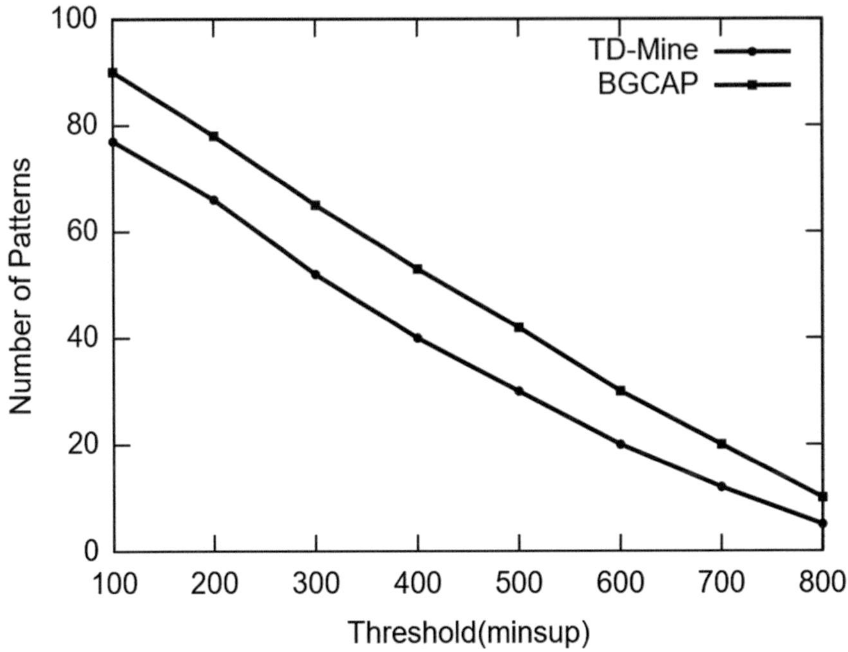

Fig. 3.10 Threshold versus number of patterns

Table 3.6 Example of prefetching rules

2-sequence pattern	Periodicity (s)	Cyclic behaviour (s)
< 1 3 >	9	57
< 1 6 >	5	93
< 3 6 >	7	134
< 1 7 >	3	68
< 6 7 >	8	74
< 5 6 >	4	101

3.6 Summary

The proposed mechanism BGCAP mines Web sequential patterns using UDDAG and analyses them using CMA to generate Web Prefetching rules. As UDDAG is based on Bidirectional pattern growth, BGCAP performs only (log n+1) levels of recursion for mining n Web sequential patterns. Prefetching rules generated based on Periodicity and Cyclic Behaviour of 2-sequence Web patterns is very accurate and the said rules hold good only until the threshold of cyclic behaviour has been reached, thus helping to implement a dynamic necessity-based prefetching strategy. Further, our experimental results show that prefetching rules generated using BGCAP is 5–10% faster for different data sizes and 10–15% faster for a fixed data size than TD-Mine. Also, BGCAP generates about 5–15% more prefetching rules than TD-Mine. The Web Prefetching rules generated from CMA are used for Caching and Prefecthing Web pages [31].

References

1. C. Jinlin, An updown directed acyclic graph approach for sequential pattern mining. IEEE Trans. Knowl. Data Eng. **22**(7), 913–928 (2010)
2. D-A. Chiang, C-T. Wang, S-P. Chen, C-C. Chen, The cyclic model analysis on sequential patterns. IEEE Trans. Knowl. Data Eng. **21**(11), 1617–1628 (2009)
3. Y. Hirate, H. Yamana, Generalized sequential pattern mining with item intervals. J. Comput. **1**(3), 51–60 (2006)
4. M.J. Zaki, Spade: an efficient algorithm for mining frequent sequences. Machine Learning, vol. 42, pp. 31–60 (2001)
5. Z. Yang, M. Kitsuregawa, LAPIN-SPAM: an improved algorithm for mining sequential pattern, in *IEEE International Conference on Data Engineering Workshops*, pp. 1222–1226 (2005)
6. Z. Yang, Y. Wang, M. Kitsuregawa, LAPIN: effective sequential pattern mining algorithms by last position induction for dense databases, in *International Conference on Database Systems for Advanced Applications*, pp. 1020–1023 (2007)
7. J. Wang, J. Han, C. Li, Frequent closed sequence mining without candidate maintenance. IEEE Trans. Knowl. Data Eng. **19**(8), 1042–1056 (2007)
8. K-Y. Huang, C-H. Chang, J-H. Tung, C-T. Ho.: COBRA: closed sequential pattern mining using bi-phase reduction approach, in *International Conference on Data Warehousing and Knowledge Discovery*, pp. 280–291 (2006)
9. J. Pei, J. Han, B. Mortazavi-Asl, J. Wang, H. Pinto, Q. Chen, U. Dayal, M-C. Hsu, Mining sequential patterns by pattern-growth: the prefixspan approach. IEEE Trans. Knowl. Data Eng. **16**(11), 1424–1440 (2004)
10. X. Cheng, H. Liu, Personalized services research based on web data mining technology, in *IEEE International Symposium on Computational Intelligence and Design*, pp. 177–180 (2009)
11. F. Lumban Gaol, Exploring the pattern of habits of users using web log sequential pattern, in *IEEE International Conference on Advances in Computing, Control and Telecommunication Technologies*, pp. 161–163 (2010)
12. J. Pei, J. Han, B. Mortazavi-asl, H. Zhu, Mining access patterns efficiently from web logs, in *Pacific-Asia Conference on Knowledge Discovery and Data Mining Current Issues and New Applications*, pp. 396–407 (2000)
13. T. Xiaoqiu, Y. Min, Z. Jianke, Mining maximal frequent access sequences based on improved WAP-tree, in *IEEE International Conference on Intelligent Systems Design and Applications*, pp. 616–620 (2006)
14. S. Yang, J. Guo, Y. Zhu, An efficient algorithm for web access pattern mining, in *Fourth International Conference on Fuzzy Systems and Knowledge Discovery*, pp. 726–729 (2007)
15. L. Liu, J. Liu, Mining web log sequential patterns with layer coded breadth-first linked WAP-tree, in *IEEE International Conference on Information Science and Management Engineering*, pp. 28–31 (2010)
16. V. Mohan, S. Vijayalakshmi, S. Suresh Raja, Mining constraint-based multidimensional frequent sequential pattern in web logs. Eur. J. Sci. Res. **36**(3), 480–490 (2009)
17. H-Y. Wu, J-J. Zhu, X-Y. Zhang, the explore of the web-based learning environment based on web sequential pattern mining, in *IEEE International Conference on Computational Intelligence and Software Engineering*, pp. 1–6 (2009)

18. B. Verma, K. Gupta, S. Panchal, R. Nigam, Single level algorithm: an improved approach for extracting user navigational patterns to technology, in *International Conference on Computer and Communication Technology*, pp. 436–441 (2010)
19. O. Nasraoui, M. Soliman, E. Saka, A. Badia, R. Germain, A web usage mining framework for mining evolving user profiles in dynamic web sites. IEEE Trans. Knowl. Data Eng. **20**(2), 202–215 (2008)
20. A. Pitman, M. Zanker, Insights from applying sequential pattern mining to E-commerce click stream data, in *IEEE International Conference on Data Mining Workshops*, pp. 967–975 (2010)
21. F. Masseglia, M. Teisseire, Pascal poncelet.: real time web usage mining with a distributed navigation analysis, in *International Workshop on Research Issues in Data Engineering*, pp. 169–174 (2002)
22. B. Zhou, S.C. Hui, K. Chang, An intelligent recommender system using sequential web access patterns, in *IEEE Conference on Cybernetics and Intelligent Systems*, pp. 393–398 (2004)
23. S-J. Yen, Y-S. Lee, M-C. Hsieh, An efficient incremental algorithm for mining web traversal patterns, in *IEEE International Conference on e-Business Engineering*, pp. 274–281 (2005)
24. Z. Zhang, X. Qian, Y. Zhao, Galois lattice for web sequential patterns mining, in *IEEE Pacific-Asia Workshop on Computational Intelligence and Industrial Application*, pp. 102–106 (2008)
25. S. Jain, R.K. Jain, R.S. Kasana, Efficient web log mining using doubly linked tree. Int. J. Comput. Sci. Inf. Sec. **3**(1), 1–5 (2009)
26. D.K. Jha, A. Rajput, M. Singh, A. Tomar, An efficient model for information gain of sequential pattern from web logs based on dynamic weight constraint, in *IEEE International Conference on Computer Information Systems and Industrial Management Applications*, pp. 518–523 (2010)
27. X. Wang, Y. Bai, Y. Li, An information retrieval method based on sequential access patterns, in *IEEE Asia-Pacific Conference on Wearable Computing Systems*, pp. 247–250 (2010)
28. K. Saxena, R. Shukla, Significant interval and frequent pattern discovery in web log data. IJCSI Int. J. Comput. Sci. Issues **7**(3), 29–36 (2010)
29. D. Oikonomopoulou, M. Rigou, S. Sirmakessis, A. Tsakalidis, Full-web prediction based on web usage mining and site topology, in *IEEE/WIC/ACM International Conference on Web Intelligence*, pp. 716–719 (2004)
30. A. Rajimol, G. Raju, Mining maximal web access patterns- a new approach. Int. J. Mach. Intell. **3**(4), 346–348 (2011)
31. K.C. Srikantaiah, N. Krishnakumar, K.R. Venugopal, L.M. Patnaik, Web caching and prefetching with cyclic model analysis of web object sequences. Int. J. Knowl. Web Intell. **5**(1), 76–103 (2014)

Chapter 4
Automatic Discovery and Ranking of Synonyms for Search Keywords in the Web

K. R. Venugopal and K. C. Srikantaiah

Abstract Accurate results for queries of keyword-based search engines are hard to come by as the queries may return irrelevant URLs even though the given keyword is present in them and some relevant URLs may be lost as they may have the synonym of the keyword and not the original one. The proposed algorithm provides solutions for these problems by making use of the Search Engine Result Pages (SERPs) to generate a ranked list of candidate synonyms for individual keywords, where the relevance of the URLs depicted by these synonyms is proved by comparing it with the URL depicted by the original keyword. This scalable technique can be applied to online data on the dynamic, domain-independent and unstructured World Wide Web. The candidate synonyms are ranked using Co-occurrence Frequencies and various page count based measures. The experimental results show that the best results are obtained using the proposed algorithm with WebJaccard.

4.1 Introduction

The World Wide Web (WWW) is a collection of interconnected Web pages accessed *via* the Internet that provides information and services from all over the world. The search engine is a Web tool that accepts query keywords as inputs, searches the keywords in its database and provides the pages that contain these keywords as Search Engine Result Pages (SERPs). Search Engines have become an indispensable part of a Web users' life as they have revolutionized the Internet usage by making

K. R. Venugopal (✉)
Bangalore University, Jnana Bharathi, Bengaluru 560056, India
e-mail: venugopalkr@gmail.com

K. C. Srikantaiah
SJB Institute of Technology, BGS Health and Education City, Uttarahalli Main Road, Kengeri, Bengaluru 560060, India
e-mail: srikantaiahkc@gmail.com

K. R. Venugopal et al., *Web Recommendations Systems*,
https://doi.org/10.1007/978-981-15-2513-1_4

tasks such as information retrieval and searching very easy and fast. Over the years, many schemes have been proposed to further enhance the features of the search engines and one such technique is the generation of synonyms for search keywords to improve the efficiency of the engine and accuracy of the search results.

Searching for the information about people is also a common activity on the Internet and it is a highly ambiguous task because a single name tends to be shared by many people. A traditional search engine such as Google and Yahoo would return Web pages in response to the search keyword entered (the person name in this case) leaving the burden of disambiguating and collecting pages relevant to a particular person (among the namesakes) on the user. A person is generally referred by multiple name aliases on the Web. Information retrieval about people from Web search engines can become difficult when a person has nicknames or name aliases. Synonyms are important for solving these problems and also various other difficulties experienced by users in the field of Natural Language Processing (NLP) such as text summarization, question answering, text generation, search query expansion, etc. Hence, it can be seen that the synonyms of search queries are quite essential.

Search engines are undoubtedly one of the best keyword-based tools for information retrieval. They generate SERPs that contain numerous links to the URLs associated with the keyword. However, one major drawback of such a mechanism is that it depends heavily on the keyword to search for the relevant URLs. In other words, (i) irrelevant URLs that contain the same keywords may inadvertently be listed among the SERPs and (ii) a page that contains information relevant to the query, but does not contain the keyword, will not be listed by the search engine (polysemy problem) and thus in both the cases, the user may not find the appropriate page. The polysemy problem may be solved if the user enters both the keyword and the synonym, but the user may not know the synonyms for all the keywords. So, all these problems can be solved by automatically extracting the synonyms from the Web.

Cheng et al. [1] proposed an offline, fully automated and data-driven algorithm called Identifying Normalization Entity Synonym that mines queries for instances where a variety of keywords have been used to refer to the same Web pages and generates an expanded set of equivalent strings called entity synonyms for each original keyword. This framework consists of three modules: candidate generation, candidate selection and noise cleaning. This technique effectively enriches the structured data with the help of these entity synonyms. This algorithm is shown to significantly increase the coverage of structured Web queries with good precision. However, it should be noted that this technique is applicable only to offline and structured data and not to the dynamic and unstructured WWW. In our approach, we solve this problem as the ASWAT algorithm is capable of working in a dynamic, online environment and it is not domain specific.

In this chapter, we propose a dynamic, online domain-independent algorithm called Automatic Discovery of Synonyms from the Web based on Inbound Anchor Text (ASWAT) that provides a ranked list of synonyms by first generating candidate synonyms. We have generated the candidates by comparing the URLs obtained in the SERPs by querying both the original keyword and its subsequent results. The key strategy of our approach is to extract the inbound anchor text as a candidate synonym when a potential match occurs on comparison of these corresponding initial and subsequent URLs. Finally, different similarity measures based on co-occurrence frequencies and page counts have been employed for ranking the candidate synonyms and further, we draw comparisons among these ranking schemes to find out the best method that gives the most relevant synonyms.

4.2 Related Works

Extensive research in the field of Web information retrieval has been achieved. However, not much effort has been put into finding people from the Web. Harada et al. [2] have proposed a method NEXAS (Named Entity extraction and Association Search) for finding authoritative people from the Web by associating a Web page through identification of its real-world entities and determines the most relevant entities considering the top-ranked Web pages from SERPs. Four simple scoring functions: document frequency (df), server frequency (sf), document frequency and inverse document frequency (dfidf) and server frequency and inverse document frequency (sfidf) are used to compute the relevance score to rank each entity. This approach is useful to see social networks reflected on the Web without explicit mentioning of relationships. However, this technique does not focus on finding other entities than people and does not take into account the co-occurrence relationships among the extracted personal names.

Kalashnikov et al. [3] proposed a system, Web Entities Search Technologies (WEST), that implements two algorithms: a Graph-based Disambiguation Algorithm for disambiguating among people who have the same name and a Graph-based Cluster Algorithm to improve Web People Search by presenting to the user a set of clusters of Web pages, one cluster per distinct person. Each cluster contains all the Web pages related to that person and allows the user to select the cluster of that person of interest. Using this method, the Web pages of the people who are not popular and which were overshadowed in the traditional search engine will be made visible to the user. However this technique does not consider external data sources such as ontologies, encyclopedia, etc., for disambiguating among people and also these algorithms are domain specific.

Since in WEST, the number of people in the shared dataset is not known in a priori, Lefever et al. [4] describe a fuzzy ant based algorithm that does not require prior specification of the number of clusters, which makes it very well suited for the Web People Search task. This technique is shown to be more robust than the agglomerative hierarchical clustering (Agnes) and k-means clustering algorithm. Balog et al. [5] proposed a Person Clustering hypothesis which states that similar documents tend to represent the same person. It is used for disambiguating a person name in a Web searching and employs two clustering techniques: Single-pass clustering (SPC) and Probabilistic Latent Semantic Analysis (PLSA).

In the Web people search, in order to improve the search quality and provide better SERPs for the user it is important to classify Web queries to know whether the query is a personal name or not. Shen et al. [6] designed an approach to predict whether a Web query is a personal name, without referring to any other context information. Personal name classification in Web queries works under two stages: offline stage and online stage. During the offline stage, probabilistic name-term dictionary is constructed for a given list of candidate name terms and during the Online stage, the probability of the query being a personal name is computed, based on the constructed Probabilistic dictionaries and some grammars. This technique outperforms the traditional approach in terms of F-score.

Jiang et al. [7] proposed the Graph-based framework for disambiguating People appearances in Web Search (GRAPE) to resolve ambiguity in the People Search. This technique first extracts People tag information such as name, organization, email address, etc., from the search results and a graph is modelled on the extracted tags. Finally, a clustering algorithm is performed on the graph to cluster all the extracted tags for each people's entity. To solve a similar problem, Smirnova et al. [8] used the link structure of the Web pages (Web graph) to disambiguate personal names using clustering algorithms.

When a person has nicknames or named aliases, the Information retrieval about that person from Web search engines can become difficult. To solve this problem, Bollegala et al. [9] devised a lexical pattern-based approach to automatically discover Personal Name Aliases from the Web. In this technique, the candidate aliases of a given name are extracted from snippets present in the SERPs of a name. These candidate aliases are ranked using three ranking approaches: lexical pattern frequency, co-occurrences in anchor texts and page counts based association measures. This technique has significantly improved the recall rate in a relation detection task and also outperforms the traditional alias extraction methods. To identify second- or higher-order associations between a name and candidates aliases, Bollegala et al. [10] proposed an approach, a Co-occurrence Graph-based approach. In this method, first an undirected co-occurrence graph is constructed. However, this approach does not extract aliases for other entity types such as products, organizations and locations.

Shen et al. [11] presented a Fuzzy set based qualitative approach model called absolute order-of-magnitude (AOM) for detecting aliases that incorporates multiple link properties such as Cardinality and Uniqueness to evaluate the similarity between information objects for the given entities and their associations. This method outperforms several methods over datasets available in the crime/terrorism-related publication and email domains. However, this technique has not been evaluated with more relevant datasets.

Finding synonyms from the Web is a challenging task as they can be associated with general terms unlike aliases that apply only to people. Kawai et al. [12] proposed a Synonym Extraction method in Specification document by considering the co-occurrence words of component words. Simanovsky et al. [13] proposed a pattern extraction algorithm to extract the text fragments between pairs of synonyms by exploiting on large-scale repositories, namely Wikipedia. In order to apply pattern extraction to Wikipedia, this technique builds a set called Synonymous phrases or markup on Wikipedia articles by considering the redirect pages titles, anchor text of backlinks and disambiguation pages titles. Next patterns are extracted, measured and ranked according to their confidence levels. Finally, these patterns can be used for extracting synonyms from the free text.

Niemi et al. [14] proposed a Bilingual Resource method for finding new synonym candidates and these are added to the existing synonym sets (synsets) of a wordnet. This technique automatically extracts groups of synonyms yielding a high number of synonyms with significant accuracy. However, the accuracy of synonym candidates, which have several possible target synsets, needs to be improved.

Plas et al. [15] describes a Distribution Similarity measure using automatic word alignment for finding synonyms using two different resources, a large monolingual corpus and a multilingual parallel corpus in eleven languages. Monolingual syntax based approaches use grammatical relations to determine the context of a target word and assume that the words that share grammatical relational contexts are semantically related. However, this approach has proven to be quite successful for finding semantically related words. Multilingual alignment based approach translates a target word into other languages found in parallel corpora and defined that as the context of target word and assumed that words that share translational contexts are semantically related. But translation will yield less semantically related words because translations do not extend to hypernyms, or hyponyms, or antonyms. Multilingual alignment based approach extracts synonyms with much greater precision and recall when compared to monolingual syntax based method.

Takeuchi et al. [16] proposed a Graph-Based Co-Clustering approach called Bipartite graph algorithm to extract verb and noun synonyms by considering co-occurrences of verbs and their arguments from large -scale texts. This method achieves a higher accuracy than those of a vector-based single clustering approach. Ageishi et al. [17] presented Statistical Machine Translation (SMT) technique to automatically extract domain specific synonyms by considering pairs of sentences from two corpora in Japanese and English languages and relies on word alignment to estimate translation probabilities.

Several similarity measures have been devised to find the synonyms and aliases. Li et al. [18] proposed a new similarity measure which is a combination of shortest path length and depth of subsuming nonlinearity to measure the semantic similarity between words. First, this method preprocesses the firsthand information from different sources. Next, words are compared within a closed interval of completely similar and nothing similar. Finally, by using this similarity measure, the appropriate semantic words are generated.

Losif et al. [19] proposed an unsupervised context-based similarity computation algorithm to compute semantic similarity between words using Web documents. This technique first downloads the top-ranked documents returned by a Web query and then computes the frequency of occurrence of contextual features. This algorithm does not refer to any external knowledge resources and can be generalized and applied to different languages. This method significantly outperforms the page count based metrics in terms of correlation scores. However, this algorithm does not take into consideration several issues such as document selection, feature selection and feature fusion.

Li et al. [20] presented a text similarity algorithm for measuring the semantic similarity between sentences or very short texts, based on semantic and word order information. Using corpus statistics and a structured lexical database, this technique computes the semantic similarity of two sentences. This method fails to disambiguate word sense using surrounding words. Han et al. [21] developed a new metric Pointwise mutual information (PMI), PMImax that augments traditional PMI to improve semantic similarity between two concepts by estimating information about the words number of senses. PMImax cannot be applied to the field of semantic acquisition from text without its combining with distributional similarity.

Bollegala et al. [22] proposed lexical pattern extraction and pattern clustering algorithms to extract the numerous semantic relations that exist between two words. First, various word co-occurrence measures are defined using page counts and then text snippets are used to extract lexical patterns. This technique outperforms the traditional semantic similarity measures by achieving a high correlation and significantly improves the accuracy in a community mining task.

Bollegala et al. [23] devised a relation extraction method based on Latent Relational Mapping that trains an existing relation extraction system (source relation) to extract new relation types (target relation) with minimum supervision. First, context in which these two entities co-occur is used to extract lexical and syntactic patterns. Next, by constructing a bipartite graph, a classifier is trained by using a small number of labeled instances to identify target relation types. It is shown that this method achieves a statistically significant F-score. But this technique does not handle unrelated entities and multiple semantic relations.

Cilibrasi et al. [24] proposed a new similarity measure called Google Similarity Distance to automatically extract similarity of words and phrases from Web using Google page counts. It is shown that this measure results in a significant mean agreement rate. Liu et al. [25] proposed Keyword Extraction algorithm using PageRank for ranking of synonyms. First a weighted co-occurrence network is used to represent the contents of a single document and in this way ranks are assigned for each synonym group using the PageRank algorithm. Lastly, several synonym groups with a high ranking are selected as keywords of the document. It is shown that this algorithm is more adaptable to different types of classes, domains or languages.

Green [26] designed a lexical chaining method for generating hypertext links both within and between documents based on semantic similarity of words and takes into account the effects of synonymy and polysemy to build Hypertext links. First the lexical chains are analysed, next similarity of paragraphs in these chains is computed. Then it decides which paragraphs should be linked based on similarities computed and connections examined among them and finally the hypertext links are built based on these connections. By using this method, the hypertext links are built with significant accuracy.

4.3 System Architecture

4.3.1 Problem Definition

Given a keyword *A*, which is a real-world entity and Web search engine *S*, our objective is to extract synonyms for a given keyword from WWW using *S* and rank candidate synonyms based on co-occurrence frequency and page count based measures. *Assumptions:* It is assumed that the user is online and only noun and verb synonyms are considered. For parent URL extraction, we take only the first five SERPs into consideration as it is assumed that only those pages contain relevant information.

Basic Definitions:

Candidate Synonyms: Synonyms of a keyword *A* are defined as the anchor texts that exactly have the same URLs linked to them as that of *A*.

Inbound Anchor Texts: It refers to a set of anchor texts that are pointing to the same URLs that are relevant to the search keyword.

The proposed architecture of our model is illustrated in Fig. 4.1 and its components are the following:

Search Engine: It is the Web tool that accepts query keywords as inputs, searches the keywords in its database and provides the pages that contain these keywords as SERPs.

Candidate Synonym Generator: It is the main component that generates candidate synonyms for the given keyword. It consists of the subcomponents: Parent URL Extractor, Child URL Extractor and Comparator. The Parent URL Extractor extracts the URLs present in the SERPs of the original query keyword. The Child URL

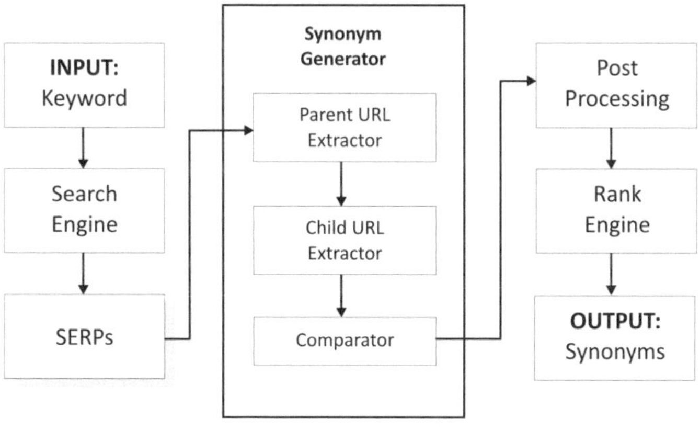

Fig. 4.1 System architecture

extractor extracts the URLs present in the SERPs that are obtained by querying the anchor texts of the parent URLs in the search engine. The comparator compares the parent URLs and the inbound anchor texts of the child URLs and if they are equal, these anchor texts are extracted as a set of candidate synonyms for the original search keyword.

Post Processing: During post-processing stage, repeating synonyms are removed and parts of speech like nouns, verbs, pronouns, prepositions, adverbs and adjectives are allotted to each word in a synonym by applying Part of Speech Tags. Only noun and verb synonyms are considered and the rest are ignored.

Rank Engine: It ranks the candidates with respect to a given keyword and identifies the correct synonym among the extracted candidate synonyms and assigns a higher rank to it. This is achieved with the help of the candidate synonyms co-occurrence frequencies and page count based measures. The end result is a ranked list of synonyms for a given keyword arranged in decreasing order of their relevance which is returned to the user.

The working of the proposed method for candidate synonym generation is briefly summarized in Fig. 4.2, which consists of the following activities:

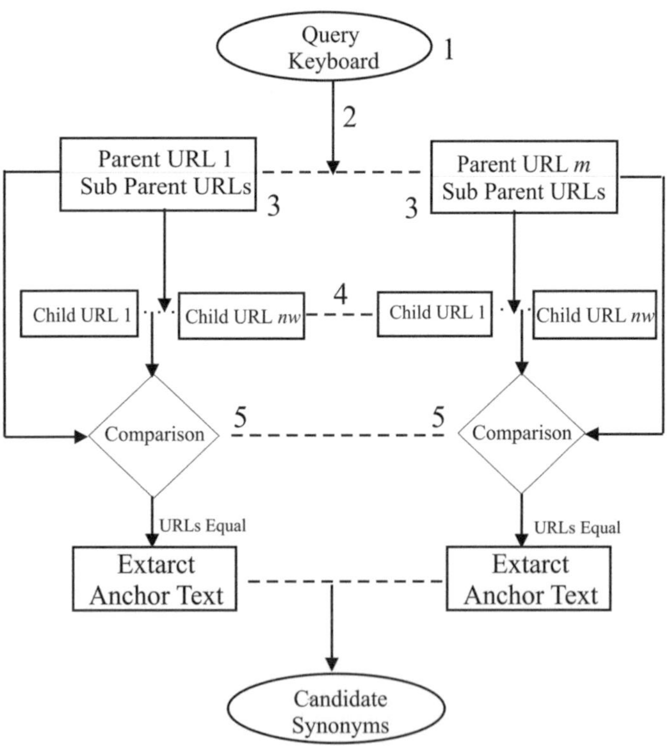

Fig. 4.2 Model flowchart

1. User inputs a keyword to the search engine to generate SERPs.
2. Parent URL Extractor extracts all the URLs contained in SERPs and is collected as a set of parent URLs.
3. For each parent URL, we retrieve the set of pages that link to parent URL from search engine and URLs contained in them are collected as subparent URLs.
4. Child URL Extractor extracts all the (anchor text, link) pairs by visiting each subparent URL and they are collected as a set of child URLs.
5. Comparator compares each of the child URLs with the parent URL and when a match occurs, the corresponding inbound anchor text of the child URL is the candidate synonym for the keyword and stored in the list.

4.4 System Model and Algorithm

4.4.1 Generation of Candidate Synonyms

First keyword A is sent as a query to the search engine S and it returns the SERPs. The URLs of all SERPs are collected to form a set of parent URLs PU, i.e.

$$PU = \{u_i \mid u_i \ is \ a \ URL \in SERP \ and \ \forall i, \ 1 \leq i \leq m\}. \quad (4.1)$$

Where m is the total number of URLs present in all SERPs for query A. Next, for each parent URL $u_i \in PU$, we retrieve the set of pages that are linked to u_i using the search engine or in other words send ui as a query to the search engine to generate the SERPs corresponding to u_i and the subsequent URLs contained in them are collected to form a set of sub parent URLs SPU, i.e.

$$SPU = \{su_{ik} \mid su_{ik} \ is \ a \ URL \in SERP \ of \ u_i \ and \ \forall k, \ 1 \leq k \leq n\}. \quad (4.2)$$

Where n is the number of URLs in SERPs of u_i. Then, for all subparent URLs $su_k \in SPU$ is visited and all the (anchor text, link) pairs that are contained in them are collected to form a set of child URLs CU_i, i.e.

$$CU_i = \{(ku_j, ka_j) \mid \forall k, \ 1 \leq k \leq n \ and \ \forall j, \ 1 \leq j \leq n\}. \quad (4.3)$$

Where w is the number of (anchor text, link) pairs retrieved for all URLs in SPU. Finally, each of these child URLs $ku_j \in CU_i$ is compared with its corresponding parent URL u_i and when $ku_j = u_j$, then the corresponding anchor text ka_j is the candidate synonym for the keyword A to form a set of candidate synonyms CS_i from URL u_i. This is represented by

$$CS_i = \{ka_j \mid ku_j = u_j, \ \forall (ku_j, ka_j) \in CU \ and \ \forall j, \ 1 \leq j \leq w\}. \quad (4.4)$$

Similarly, for the remaining URLs in PU, the candidate synonyms can be generated. The candidate synonyms CS for the given keyword A is the Union of candidate synonyms $CS_1, CS_2, \ldots CS_m$, i.e.

$$CS = CS_1 \cup CS_2 \cup \cdots \cup CS_m \tag{4.5}$$

Once a set of candidate synonyms are extracted, a part of speech like noun, verb, pronoun, preposition, adverb, adjective, etc., is assigned to each word in a synonym by applying Part of Speech (PoS) tagger and only noun and verb synonyms are considered.

Example 4.1 A keyword $A = Fireblade$, which is a sports motorcycle manufactured by Honda, is entered as a query in the Google search engine to obtain the SERPs. The URLs contained in the SERPs are collected as a set of parent URLs, i.e. $PU = \{http://en.wikipedia.org/wiki/Honda_Fireblade, http://world.honda.com/CBR1000RR/ ...\}$.

For the sake of notation, we shall refer to the URLs in PU as $u_1, u_2 \ldots$ that are separately queried in the search engine to generate the SERPs and the new URLs contained in them are collected as sub parent URLs, (SPU_1 from u_1, SPU_2 from $u_2, \ldots$) $SPU_1 = \{http://en.wikipedia.org/wiki/2008_Isle_of_Man_TT, http://en.wikipedia.org/wiki/2003_Superbike_World_Championship_season, ...\}$, $SPU_2 = \{http://world.honda.com/siteindex/, ...\}$.

By visiting all sub parent URLs that belongs to the set SPU_1, SPU_2... the (anchor text, hyperlink) pairs that are contained in them are collected as a set of child URLs CU_1, CU_2..., respectively, i.e. $CU_1 = \{(http://en.wikipedia.org/wiki/Honda_Fireblade$, Honda Fireblade), $(http://en.wikipedia.org/wiki/Honda_Fireblade$, Honda CBR954RR), $(http://en.wikipedia.org/wiki/Honda_RC51$, Honda VTR 1000SPW) } CU_2= {($http://world.honda.com/CBR1000RR/$, CBR1000RR) }.

Ultimately u_1 is compared with the URL part of the elements of CU_1 and u_2 is compared with the URL part of the elements of CU_2 and so on. Since u_1 is the same as the URL part of the first element of CU_1, the corresponding anchor text part Honda Fireblade is one of the candidate synonyms. Similarly, this process is repeated for the entire set of parent URLs and that synonyms are stored as a list, i.e. CS_1 = { Honda Fireblade, Honda CBR954RR } CS_2 ={ CBR1000RR }

This process is repeated for the entire set of parent URLs. Ultimately CS contains Honda Fireblade, Honda CBR954RR and CBR1000RR as candidate synonyms for the keyword $Fireblade$, i.e. CS = {Honda Fireblade, Honda CBR954RR and CBR1000RR}.

4.4.2 Ranking of Candidate Synonyms

For a given keyword A, many candidate synonyms are generated and the most relevant among these candidates must be ranked in the first position. This ranking can be achieved by finding the association between a keyword and the candidate synonyms. To compute this association, two techniques: co-occurrence frequency (CF) and page counts based measures, can be employed to rank the candidate synonyms.

Co-occurrence Frequency (CF): CF between a unique keyword A and its candidate synonym $c_i \in CS$ is the number of distinct URLs that are linked to A and c_i. The higher the CF, the more relevant the candidate synonym and so the CF is computed between A and all its candidate synonyms and ranked based on the decreasing order of the CFs. Hence the candidate with the highest CF value will be ranked in the first position and so on.

Example 4.2 CFs between the keyword $Fireblade$ and its candidate synonyms generated in Example 4.1 are: (Fireblade, Honda Fireblade) is 3, (Fireblade, Honda CBR954RR) is 2 and (Fireblade, CBR1000RR) is also 1. As (Fireblade, Honda Fireblade) has the highest CF value, Honda Fireblade is the most relevant synonym for the keyword $Fireblade$ and ranked in the first position and the other two synonyms are ranked in the second and third positions, respectively.

Page Count based Measures: The Page Count of a query keyword A is an estimate of the number of pages that contain the query keyword. The candidate synonyms are ranked using similarity measures that in turn use page counts of A only, c_i only and both A and c_i. Let N_A be the page counts for keyword A only, N_{c_i} be the page counts for a candidate synonym c_i only and N_{Ac_i} be the page counts for both keyword A and a candidate synonym c_i.

A similarity measure is a function which computes the degree of similarity and represents the similarity between two words. We use nine popular relevance measures: WebJaccard, Cosine, WebDice, WebOverlap, Precision, Recall, WebPMI (Point-wise Mutual Information) and NGD (Normalized Google Distance), to accurately measure the relevance between A and c_i and to rank the candidate synonyms using page counts. These similarity measures are described using N_A, N_{c_i} and N_{Ac_i} as follows.

WebJaccard Coefficient: is the measure used for comparing the similarity and diversity of sample sets and is defines as the size of the intersection divided by the size of the union of the sample sets. WebJaccard Coefficient between keyword A and a candidate synonym $c_i \in CS$ is defined as

$$WebJaccard(A, c_i) = \frac{N_{Ac_i}}{N_A + N_{c_i} - N_{Ac_i}}. \tag{4.6}$$

Cosine Similarity: It is a measure of similarity between two vectors that measures the cosine of the angle between them. Cosine Similarity between keyword A and a candidate synonym $c_i \in CS$ is defined as

$$cos(A, c_i) = \frac{N_{Ac_i}}{\sqrt{N_A}\sqrt{N_{c_i}}}. \tag{4.7}$$

WebDice (A, c_i): between keyword A and a candidate synonym $c_i \in CS$ is defined as

$$WebDice(A, c_i) = \frac{2N_{Ac_i}}{N_A + N_{c_i}}. \tag{4.8}$$

WebOverlap (A, c_i): between keyword A and a candidate synonym $c_i \in CS$ is defined as

$$Overlap(A, c_i) = \frac{N_{Ac_i}}{min(N_A + N_{c_i})}. \tag{4.9}$$

F-score: It can also be used to rank the candidate synonyms. F-score of a synonym c_i is computed as the harmonic mean between the precision and recall of the synonym. First, for a synonym c_i and a keyword A, we compute its precision and recall as follows: Precision is the ratio between the number of relevant documents that are returned and the total number of returned documents. Precision (A, c_i) between A and c_i is defined as

$$Precision(A, c_i) = \frac{N_{Ac_i}}{N_A}. \tag{4.10}$$

Recall is the ratio between the number of relevant documents that are returned and the total number of relevant documents. Recall (A, c_i) between A and c_i is defined as

$$Recall(A, c_i) = \frac{N_{Ac_i}}{N_{c_i}}. \tag{4.11}$$

Then, its F-score is computed as

$$F - score = \frac{2 * Precision(A, c_i) * Recall(A, c_i)}{Precision(A, c_i) * Recall(A, c_i)}. \tag{4.12}$$

WebPMI between keyword A and a candidate synonym $c_i \in CS$ is defined as

$$WesPMI(A, c_i) = log_2\left(\frac{LN_{Ac_i}}{N_A N_{c_i}}\right). \tag{4.13}$$

Normalized Google Distance (NGD): is a semantic similarity measure derived using page counts retrieved from Google search engine. The lesser the NGD between two words, higher the similarity and vice versa. If L is the number of documents indexed by the search engine, then NGD (A, c_i) between keyword A and a candidate synonym $c_i \in CS$ is defined as

$$NGD(A, c_i) = \frac{max(logN_A, logN_{c_i}) - logN_{Ac_i}}{logL - max(logN_A, logN_{c_i})}. \tag{4.14}$$

Algorithm 4.1: ASWAT Algorithm.

```
Input  : Keyword A.
Output : List of Synonyms for A, ranked in the order of relevance.
//Generation of Candidate Synonyms
SERPs = SearchEngine(A)
PU = ExtractAllURLs(SERPs)
SPU=∅
CS=∅
for each URL u_j ∈ PU do
    SPU_i = get all pages that contains URL u_j
    SPU =∪i SPU_i
    CU_i=∅
    for each URL su_k ∈ SPU do
        P = Visit the Web page corresponding to URL su_k
        CU_ik = collect all anchortext and URL pair (ku_j, ka_j) from P;
        CU_ik =∪ik CU_ik
    for each ku_j ∈ CU do
        If (u_i == ku_j) CS_i =∪i ka_j
    CS =∪i CS_i
//Ranking Candidate Synonyms
for each URL c_i ∈ CS do do
    Calculate co-occurrence frequency A and c_i
    Compute the page counts for A (N_A)
    Compute the page counts for A and c_i (N_Ac_i)
    Compute the page counts for c_i (N_c_i)
    Rank candidate synonym c_i ∈ CS
    WebJaccard(A,c_i) = N_Ac_i / ( N_A+ N_c_i - N_Ac_i )
    Cos(A,c_i) = N_Ac_i / Sqrt(N A) Sqrt(N_c_i)
    WebDice(A,c_i) = (2 N_Ac_i ) / ( N_c_i + N_A )
    WebOverlap(A,c_i) = N_Ac_i / min(N_A , N_c_i)
    Precision(A,c_i) = N_Ac_i / N_A
    Recall(A,c_i) = N_Ac_i / N_c_i
    F-Score =(2Precision(A,c_i)Recall(A,c_i)) /(Precision(A,c_i) + Recall(A,c_i) )
    WebPMI(A,c_i) = log 2 ( L N_Ac_i / N_A N_c_i)
    NGD(A,c_i) = ( maxlog N_A, log N_c_i log N_Ac_i ) / (log L minlog N_A, log N_c_i )
```

4.4.3 ASWAT Algorithm

The Automatic Discovery of Synonyms from the Web based on Inbound Anchor Text (ASWAT) algorithm is shown in Algorithm 4.1, which extracts synonyms from the Web and is based on the described model. In this algorithm, the search engine first uses the keyword to generate the SERPs and the URLs contained in them are collected as a set of parent URLs. A hash map that stores the URLs of the pages that contain links to these *parent URLs* is maintained as SPU. Next, each page of SPU

is visited and all the (anchor text, link) pairs that are contained in it are collected as a set of *child URLs* for this parent. Finally, each of these child URLs is compared with the parent URL and when a match occurs, the inbound anchor text of the child URL is extracted and stored in the hash map with the parent URL. This process is repeated for the entire set of parent URLs and ultimately the hash map contains candidates for synonyms for the keyword. The next stage of the algorithm is to rank the candidate synonyms based on the co-occurrence frequency and page count based measures.

4.5 Experiments

The ASWAT algorithm has been implemented using Java language using the NetBeans 7.1 IDE and Google search engine in a Pentium Dual-Core processor environment, with 2GB of RAM and 100 GB HDD. Experiments were conducted to derive the candidate synonyms for some popular search keywords such as titles of movies, camera models, motorcycle models and place names. For parent URL extraction, only the first five SERPs have been considered and for the subparent and child URLs, all the SERPs have been taken into consideration.

Table 4.1 lists the query keywords used and their synonyms along with their CFs. The synonyms with the highest CFs are listed first and are hence the most relevant

Table 4.1 Common keywords and their synonyms

Keywords	Candidate synonyms (Co-occurrence frequency)
Danny the Dog	Unleashed (4), Unleashed (aka Danny the Dog) (2), Unleashed/Danny the Dog (1), Danny the Dog (2005) (1)
Bengalooru	Bangalore (2), Bengaluru (1), Garden City (1) IT Capital of India (1)
Fireblade	Honda Fireblade (3), CBR1000RR (1) Honda CBR1000RR (1)
Live Free or Die Hard	Die Hard 4 (3), Die Hard 4.0 (2) Live Free or Die Hard (2007) (2), Die Hard IV (1)
Armour of God	Operation Cordor 2: The armour of God (1) Long Xiong Hu Di (1), Armour of God (film) (1)
Welcome to the Jungle	The Rundown (1), The Rundown (Welcome to the Jungle) (1) Guns N' Roses -Welcome to the Jungle (1)
Canon EOS 400D	Canon EOS Kiss Digital X (1), EOS Rebel XTi (1) 400D Digital Rebel XTi Kiss Digital X (1)
Republic of India	India (2), Bharat (1)
US	USA (2), United States of America (1)
Mac OS X	Leopard (2)

Table 4.2 Ranked synonyms for *Begalooru*

Synonym	WebJaccard	Cosine	WebDice	WebOverlap	F-score	NGD
Bangalore	0.1275	0.396	0.2261	1.263	0.2262	0.7098
Bengaluru	0.0141	0.1188	0.0279	1	0.0279	0.6598
Garden City	0.0029	0.0535	0.0057	1.0094	0.0058	0.4541
IT Capital of India	0.0073	0.0145	0.0144	0.0154	0.0145	0.1387

Table 4.3 Ranked synonyms for *Armour of God*

Synonym	Web Jaccard	Cosine	WebDice	WebOverlap	F-score	NGD
Armour of God (film)	0.0088	0.1188	4.0262	6.8824	1.1327	0.6462
Operation Condor 2: The Armour of the Gods	0.0098	0.0535	0.0115	0.8806	0.0154	0.6289
Long Xiong Hu Di	0.0123	0.0145	0.0476	0.1903	0.0244	0.5919

synonyms associated with those keywords. Some synonyms have the same CF value which implies that their relevance is the same and can be listed in any order.

Tables 4.2 and 4.3 specifically show the synonyms for the keywords, *Bengalooru* and *Armour of God*, respectively. Here, the page count based measures have been used to evaluate the similarity between the keywords and their synonyms to compare the ranking of these synonyms under these measures. From Table 4.2, it can be observed that the synonym *Bangalore* has the highest similarity score compared to the other synonyms with respect to all the measures mentioned and hence it is the most relevant synonym for the keyword *Bengalooru*. *Bengaluru* has the second highest similarity score and *IT Capital of India* the least. Similarly, from Table 4.3, the synonym *Armour of God (film)* has the highest similarity score for the keyword *Armour of God*.

To compare the similarity measures, we have used Mean Reciprocal Rank (MRR) [9]. If r_i is the rank of the correct synonym c_i and n is the total number of candidate synonyms, then MRR is defined as

$$MRR = \frac{1}{n}\sum_{i=1}^{n}\frac{1}{r_i}. \tag{4.15}$$

Table 4.4 lists the MRR in descending order for the keywords *Bengalooru* and *Armour of God* for different measures. It can be observed that correct synonyms are obtained by using the proposed algorithm with WebJaccard.

Table 4.4 MRR scores for page count based measures

Measure	MRR (Begalooru)	MRR (Armour of God)
WebJaccard	70.07238	37.12225
WebDice	35.64352	13.52666
F-score	35.20531	13.35019
Cosine	12.32498	12.00933
WebOverlap	8.46469	0.816969
NGD	1.542053	0.603383

4.6 Summary

The ASWAT algorithm, after effective scanning and processing, provided a ranked list of candidate synonyms for each search keyword. This new technique is scalable and can be implemented in the unstructured and dynamic Web i.e. it can be applied on online and domain-independent content. This mechanism not only provides accurate synonyms, but also mitigates the polysemy problem to a large extent thus providing the user with valuable and correct links for the desired information. The candidate synonyms are ranked using CFs and various page count measures. The experimental results have shown that correct synonyms were obtained using the proposed algorithm with WebJaccard. As a future enhancement to ASWAT, snippets in the SERPs can also be taken into consideration for generating synonyms.

References

1. T. Cheng, H.W. Lauw, S. Paparizos, Entity synonyms for structured web search. IEEE Trans. Knowl. Data Eng. **24**(10), 1862–1873 (2012)
2. M. Harada, S. Sato, K. Kazama, Finding authoritative people from the web, in *Proceedings of the Joint ACM/IEEE Conference on Digital Libraries (JCDL04)* (2004), pp. 306–313
3. D.V. Dmitri, Kalashnikov, Z. Chen (Stella), S. Mehrotra, R. Nuray-Turan, Web people search *via* connection analysis. IEEE Trans. Knowl. Data Eng. **20**(11), 1–16 (2008)
4. E. Lefever, T. Fayruzov, V. Hoste, M. De Cock, Fuzzy ants clustering for web people search, in *2nd Web People Search Evaluation Workshop (WePS 2009), Madrid, Spain* (2009)
5. K. Balog , L. Azzopardi, L. Azzopardi, Personal name resolution of web people search, in *NLPIX2008, Beijing, China*, 22 April 2008
6. D. Shen, T. Walker, Z. Zheng, Q. Yang, Y. Li, Personal name classification in web queries, in WSDM 08, Palo Alto, California, USA (2008), pp. 149–158
7. L. Jiang, J. Wang, N. An, S. Wang, J. Zhan, L. Li, GRAPH: a graph-based framework for disambiguating people appearances in web search, in *9th IEEE International Conference on Data Mining* (2009), pp. 199–208
8. E. Smirnova, K. Avrachenkov, B. Trousse, Using web graph structure for person name disambiguation, in *3rd Web People Search Evaluation Forum (WePS-3), CLEF Conference* (2010)
9. D. Bollegala, Y. Matsuo, M. Ishizuka, A co-occurrence graph-based approach for personal name alias extraction from anchor texts, in *International Joint Conference on Natural Language Processing* (2008), pp. 865–870

10. D. Bollegala, Y. Matsuo, M. Ishizuka, Automatic discovery of personal name aliases from the web. IEEE Trans. Knowl. Data Eng. **23**(6), 831–844 (2011)
11. Q. Shen, T. Boongoen, Fuzzy orders-of-magnitude-based link analysis for qualitative alias detection. IEEE Trans. Knowl. Data Eng. **24**(4), 649–662 (2012)
12. Y. Kawai, T. Yoshikawa, T. Furuhashi, E. Hiraoy, A. Kunoy, T. Gotohy, A study on extraction method of synonyms in specification documents, in *13th ACIS International Conference on Software Engineering, Artificial Intelligence, Networking and Parallel/Distributed Computing* (2012), pp. 321–324
13. A. Simanovsky, A. Ulanov, Mining text patterns for synonyms extraction, in *22nd International Workshop on Database and Expert Systems Applications* (2011), pp. 473–477
14. J. Niemi, K. Linden, M. Hyvarinen, Using a bilingual resource to add synonyms to a wordnet: FinnWordNet and wikipedia as an example, in *6th International Global Wordnet Conference, Matsue Japan* (2012), pp. 227–231
15. L. van der Plas, J. Tiedemann, Finding synonyms using automatic word alignment and measures of distributional similarity, in *Annual Meeting of the Association of Computational Linguistics* (2006), pp. 866–873
16. K. Takeuchi, Extraction of verb synonyms using co-clustering approach, in *2nd International Symposium on Universal Communication* (2008), pp. 173–178
17. R. Ageishi, T. Miura, Automatic extraction of synonyms based on statical machine translation, in *22nd International Conference on Tools with Artificial Intelligence* (2010), pp. 313–317
18. Y. Li, Z.A. Bandar, D. McLean, An approach for measuring semantic similarity between words using multiple information sources. IEEE Trans. Knowl. Data Eng. **15**(4), 871–882 (2003)
19. E. Iosif, A. Potamianos, Unsupervised semantic similarity computation between terms using web documents. IEEE Trans. Knowl. Data Eng. **22**(11), 1637–1647 (2010)
20. Y. Li, D. McLean, Z.A. Bandar, J.D. OShea, K. Crockett, Sentence similarity based on semantic nets and corpus statistics. IEEE Trans. Knowl. Data Eng. **18**(8), 1138–1150 (2006)
21. L. Han, T. Finin, P. McNamee, A. Joshi, Y. Yesha, Improving word similarity by augmenting PMI with estimates of word polysemy. IEEE Trans. Knowl. Data Eng. **25**(6), 1307–1322 (2013)
22. D. Bollegala, Y. Matsuo, M. Ishizuka, A web search engine-based approach to measure semantic similarity between words. IEEE Trans. Knowl. Data Eng. **23**(7), 977–990 (2011)
23. D. Bollegala, Y. Matsuo, M. Ishizuka, Minimally supervised novel relation extraction using a latent relational mapping. IEEE Trans. Knowl. Data Eng. **25**(2), 419–432 (2013)
24. R.L. Cilibrasi, P.M.B. Vitanyi, The google similarity distance. IEEE Trans. Knowl. Data Eng. **19**(3), 370–383 (2007)
25. Z. Liu, J. Liu, W. Yao, C. Wang, Keyword extraction using PageRank on synonym networks, in *ICEEE, Henan* (2010), pp. 1–4
26. S.J. Green, Building hypertext links by computing semantic similarity. IEEE Trans. Knowl. Data Eng. **11**(5), 713–730 (1999)

Chapter 5
Construction of Topic Directories Using Levenshtein Similarity Weight

K. R. Venugopal and K. C. Srikantaiah

Abstract Topic directories are search engines consisting of categories in hierarchical manner. Mapping of a new Web page to an appropriate category of a topic directory is one of the major challenges faced by human-based topic directories due to the rapid pace of growth of the WWW and also the presence of a large number of categories. So, the mapping of new pages onto categories by human experts is an expensive process. Hence, the automation of this process is needed and can be performed using standard similarity measures. In this chapter, we propose an algorithm called Mapping of Web Pages to Categories using Levenshtein Similarity Weight (MPC-LSW algorithm) that performs this mapping of Web pages to categories by comparing the similarity of the pages, i.e. a page-based comparison instead of the traditional term-based comparison. The time complexity of MPC-LSW is observed to be O(mk) as the terms are eliminated and processing is faster because pages are compressed into strings. Hence, it is an efficient method of mapping.

5.1 Introduction

The World Wide Web (WWW) is a collection of interconnected Web pages accessed via the Internet offering information and data from all over the world. A topic directory, also known as type-oriented Web directory is a highly organized structure of Web pages that supports semantics-based information searches. Consider the example of a university website, where searches to the website can be made more efficient by using a topic directory like *India, Universities, Bangalore University, Department of CSE,* etc., which make such directories highly desirable. But, these directories must

K. R. Venugopal (✉)
Bangalore University, Jnana Bharathi,
Bengaluru 560056, India
e-mail: venugopalkr@gmail.com

K. C. Srikantaiah
SJB Institute of Technology, BGS Health and Education City,
Uttarahalli Main Road, Bengaluru 560060, India
e-mail: srikantaiahkc@gmail.com

K. R. Venugopal et al., *Web Recommendations Systems*,
https://doi.org/10.1007/978-981-15-2513-1_5

be constructed manually by the human experts and hence they provide only limited coverage with no regard to scalability and adaptability. Also, when searching for a topic in the WWW using topic directories, it returns many links or Web sites related to a given topic and each topic falls under distinguished categories like Sports, Politics, Health,.. etc. Each searched topic should necessarily be mapped to an existing category based on its similarity.

Currently, Web directories like Yahoo! make use of human readers to classify Web documents which may have major ramifications with respect to the cost and speed of the operations. Hence there is a demand for making the mapping process automatic so as to enhance these factors to have high speed and accuracy at a reduced cost. Although several automatic classification techniques are currently being implemented, none of them seem to improve the factor of speed even though there is a significant improvement in accuracy and cost reduction. In this chapter, we address all these factors, especially speed, by eliminating the term-based comparison for mapping the given Web pages onto their respective topic directories.

There are two types of Web directories: $artificial$ and $real$. Real Web directories are those that are constructed by human experts to map newly created authentic Web pages corresponding to a certain topic. Artificial Web directories are those that are constructed using the pages present in the Web log. Since the records available in the Web logs is vast, new pages viewed by the user should be added to the artificial Web directory and this can be utilized for constructing community Web directories, thus personalizing these directories. This is achieved by employing one of the mapping techniques based on similarity observed between the new page and categories in the artificial Web directory.

Pierrakos et al. [1] have used cosine similarity function for mapping of browsed Web pages onto categories of Web directory. Here, Term Frequency and Inverse Document Frequency (TF-IDF) is calculated for all the terms in the Web pages and also for terms in the categories. Then cosine similarity is applied between the Web page and the category. The time complexity of this approach, i.e. mapping the Web pages to categories is $O(mkn)$, where m is the number of categories, k is the number of Web pages in each category and n represents the number of terms.

In this chapter, we have proposed an efficient algorithm called Mapping of Web Pages to Categories using Levenshtein Similarity Weight (MPC-LSW) that performs mapping of Web pages to categories by first compressing the page to be mapped and also the pages in the category into strings using Hashing technique. Next, the Levenshtein Distance (LD) between all possible pairs of the string of the given page and each string corresponding to every existing page in the category is computed. Finally, Levenshtein Similarity Weight (LSW) is found to map this given page to the appropriate category automatically.

5.2 Related Works

Hyyro [2] designed an efficient algorithm to find LD based on bit vector. Burkhardt et al. [3] proposed a filtering algorithm based on one gapped q-Gram for preprocessing the input strings. LD is being used for a variety of applications such as the establishment of secure communication between light client and single server [4], information integration from different heterogeneous systems [5], for correcting output words of Optical Character Recognition using thematic Web pages from Dynamic Web Directories [6], analysing simultaneous speech recognition systems automatically [7], to find coherent chunks of text in the document for text ranking [8], a similarity measure to cluster Web logs and finally as a technique to find the dissimilarity between Web pages at the content level, which is used to cluster the Web pages [9, 10].

Web page classification is different from normal text categorization because the Web page contains different types of data such as text, hyperlinks, media and noise. The following techniques discuss the classification of the existing Web pages into many categories. Fathy [11] deigned the Automatic Information Retrieval Database system (AIRDB) to classify Web pages based on Artificial Neural Networks and stemming. Classification is done using keyword indexes and weight of each stem which results in a classification performance of about 84%. However, AIRDB is not capable of classifying newly added pages. Soonthornphisaj et al. [12] proposed the Iterative Cross Training (ICT) which is a semi-supervised algorithm that utilizes Inductive Logic Programming and two learners, strong and weak. ICT induces the ability of the weak learner by finding a strong learner and consequently performs the classification in the real world by making use of the information provided by this strong learner. ICT is shown to outperform supervised Naive Bayes, Co-training and Expectation–Maximization Algorithm. However, there is scope for improving the performance of the weak learner.

Shen et al. [13] proposed a new automatic Web page summarization algorithm that utilizes a page-layout analysis to obtain the important topic of the Web page. This technique is shown to enhance the accuracy of the classification. It combines the results of four summarization mechanisms for Web page classification, viz. Adapted Luhn's Summarization, Latent Semantic Analysis, Content Body Identification by Page Layout Analysis and Naive Bayes Summarization. This algorithm is shown to achieve an improvement of about 12.9% compared to pure-text based methods. It outperforms summary generated by human editors by 14.8%. However, hyperlinked Web pages are not being considered for summarization of multiple documents. Lin et al. [14] surveyed and compared the different Web page classification techniques and listed their issues.

Riboni [15] designed an efficient way for the representation of the Web pages, which combines usual Vector Space Representation Method and Co-matrix for classification which is shown to improve the classification performance. Lakshminarayana [16] proposed a categorization algorithm that utilizes two different Web page weights, viz. Page Retaining Weight (PRW) and Page Forwarding Weight (PFW) that improves the retrieval performance of the search engine. Javid et al. [17] used the swarm intelligence algorithm for Web page classification. Bo et al. [18] used the traditional techniques, viz. Hidden Naive Bayes (HNB), Naive Bayes, SVM, k Nearest Neighbour, C 4.5, Complement Class Naive Bayes for classification and concluded that HNB is better than the rest.

Kim et al. [19] proposed a semi-supervised learning method for Web page classification that leverages a certain hypothesis and augments the training set by modelling the similarity between documents in a click graph. The hypothesis is that unlabelled documents similar to positive and negative labelled documents tend to be clicked through by the same user queries. This expands training data automatically by finding similar pages from user click information. Further two models are proposed for seed set augmentation: Click Data Model (CDM) and Query Constraint model (QC model). In CDM, a page with high similarity with the seed set should have the same class label as the seed set. In the QC model, an active learning method leveraging click data attempts to reduce the number of labelled data required to build a model of similar performance by actively collecting training data. This technique is shown to outperform current models with less amount of human annotation effort. However, the quality of automatically labelled data from unlabelled data needs improvement.

Prakash et al. [20] proposed the technique for the automatic classification of the Web pages into broad categories based on the structure and images pertaining to each Web page. This structure-based classification approach exploits the fact that Web pages belonging to a particular category all have a similar structure and also that there are many other features such as hyperlinks and media other than the text content. The technique consists of feature extraction, which gathers textual information and image information for classification. The technique is shown to achieve an accuracy of 87.83% for correctly classified Web pages. However, only text and images are being considered for classification and other information in the form of video, audio, etc. are not being considered for exploitation.

Pierre [21] proposed two techniques called Targeted Spidering and Opportunistic Crawling for automated text classification of Web sites by analysing the use of HTML meta-tags as a good source of text features. It is shown that HTML meta-tags are not in wide use despite their role in search engine rankings for automatically classifying websites into industry categories. The text classifier consists of three modules: the targeted spider for extracting text features associated with a website, an information retrieval engine for comparing queries to training examples and a decision algorithm for assigning categories. This method is shown to be reliable and accurately generate metadata from existing content. However, creation of quality metadata is a tedious process.

Yu et al. [22] proposed a heterogeneous learning framework for classifying Web pages, by introducing two heterogeneous learners: a decision list and a linear separator which complement each other. This technique increases classification accuracy and eliminates the need for negative training data. It is shown that this classifier classifies more accurately while requiring less human efforts.

Yu et al. [23] presented a framework called Positive Example Based Learning (PEBL) for Web page classification which eliminates the need for manually collecting negative training examples in preprocessing. PEBL applies an algorithm called Mapping-Convergence (M-C) to achieve high classification accuracy (with positive and unlabeled data) as high as that of a traditional SVM (with positive and negative data). M-C runs in two stages: the mapping stage (where a weak classifier draws an initial approximation of strong negative data) and the convergence stage that iteratively runs an internal classifier (e.g. SVM) which maximizes margins to progressively improve the approximation of negative data. It is shown that the M-C algorithm outperforms one-class SVMs and is almost as accurate as the traditional SVMs. However, the M-C algorithm takes multiple times longer to train one class than traditional SVM since the iteration of training SVM has to be serialized due to data dependency between adjacent iterations.

Prabowo et al. [24] designed an automatic classifier which utilizes ontologies with respect to the Dewey Decimal Classification (DDC) and Library of Congress Classification (LCC) schemes. This technique first builds these ontologies in a modular fashion. This is achieved by a strategy used to enhance the accuracy of the automatic classifier, called the Automatic Classification Engine (ACE). The advantages of an ontology-based classification over hierarchical and probabilistic approaches are: enabling of machine reasoning, inherent semantics and a new way to measure the similarity between a Web page and a class representative. This technique is also shown to result in improved classification in terms of accuracy. However, due to the incompleteness of the ontologies used, it results in a low coverage ratio.

Several techniques have been proposed for mapping a Web page onto a category discussed as follows. Tsukada et al. [25] proposed techniques to generate attributes by using co-occurrence analysis and to classify Web pages automatically based on machine learning. By applying these techniques to Web pages, decision trees are constructed which determine appropriate category for each Web page. The performance of this proposed method is evaluated in terms of error rate, recall and precision and is shown to yield high accuracy with the classification values of Precision are higher than those of Recall for each category. The values of Error rate is between 8 and 16%. However, for the evaluation only the top-level category classification has been considered.

Patil et al. [26] proposed a technique which makes use of the Naive Bayesian (NB) approach for the classification of Web sites based on content of the home pages based on the fact that the expression power of the home page of a website can be exploited to identify the nature of the organization. Using this technique yields an average precision of 89.09%, average recall of 89.04%, and F-Measure of 89.05% and it is shown that classification of Web sites is possible by examining the contents of their home pages. However, in this technique, only distinct and non-hierarchical categories are considered.

Peng et al. [27] proposed a dynamic and hierarchical classification system that is capable of adding new categories as required for the organization of the Web pages into a tree structure. Classification of Web pages is achieved by searching through only one path of the tree structure. As the number of Web pages in the WWW is increasing continuously at great speed, it is impossible for a fixed category set to provide accurate classification and hence this dynamic expanding technique is most advantageous. It is shown that the single-path search technique reduces the search complexity and increases the accuracy by 6% comparing to related algorithms.

The problem of grouping similar Web pages can be done by using Clustering. Chau et al. [28] proposed the use of Web structure and hyperlink analysis techniques in traditional vector space model for clustering of Web documents returned by search engines. It introduces a new metric HFIDF (Hyperlink Frequency multiplied by Inverse Document Frequency) instead of traditional TF-IDF. This technique exploits the fact that a document is represented as a vector based on the vector space model in document clustering. HFIDF can be used to evaluate several clustering algorithms such as K-means clustering, self-organizing maps (SOM), Suffix-Tree clustering and agglomerative hierarchical clustering. Based on the standard clusters created manually, the performance of HFIDF and TFIDF is compared in terms of cluster Precision and cluster Recall.

Lin et al. [29] proposed a clustering algorithm HSClus which groups densely linked Web pages into semantic clusters and identifies their hierarchical relationships by studying the DOM structure of the pages to extract parallel links that depict similarity amongst the pages. Here, both the cross-page link structure and in-page link references are used to identify their hierarchical relationships. It is shown that HSClus is an efficient and effective technique for discovering Web page structures of websites for hierarchical clustering by utilizing the inherent semantic structure of the website. Shi [30] proposed a rough k-means clustering algorithm that utilizes fuzzy logic and Web usage mining to cluster Web pages based on their browsing patterns.

5.3 System Architecture

5.3.1 Problem Definition

Given a Web log that contains URLs of the visited Web pages and a set of categories C, our objective is to download each page, compress using hashing, compute similarity using LD and map them onto categories using Levenshtein Similarity Weight(LSW).

Assumptions: In this chapter, we have assumed that the text portion of the Web pages is most significant for mapping and hence only text has been taken into consideration. We have also assumed that Web pages are always present in the categories, i.e. a category must consist of at least one page for mapping and a browsed page must be mapped to at least one category.

The entire functionality is depicted in Fig. 5.1, which consists of the following activities:

1. Interested Web pages being browsed by the user.
2. Address of the Web page which is stored in the Web log and corresponding pages are retrieved using crawler.
3. Extracting categories from the Web directory.

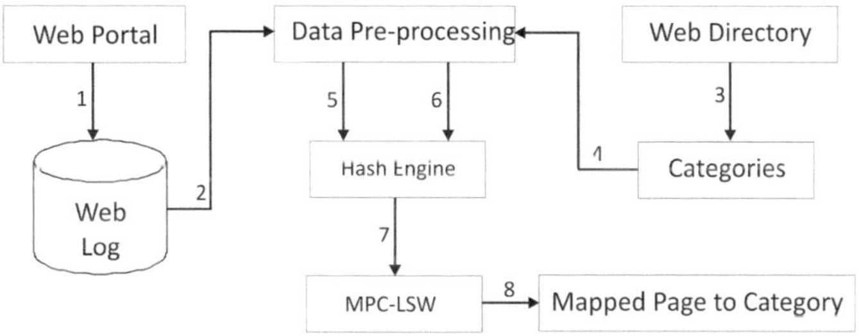

Fig. 5.1 System architecture

4. Extracting pages present in category.
5. Pages obtained from Web log are preprocessed and forwarded to Hash Engine.
6. Pages obtained from categories are preprocessed and forwarded to Hash Engine.
7. Hashed values for corresponding to the browsed pages and pages in categories.
8. Mapping of pages to categories based on similarity is performed.

Web Portal: A Web portal is a website that brings information together uniformly from a variety of sources. The main idea is to provide the Web user with a single Web page that brings together or aggregates content from a number of other systems or servers. Each information source gets its dedicated area on the page for displaying information and the Web user can configure which ones to display. Apart from the standard search engines features, Web portals offer other services such as email, news, stock prices, information, databases and entertainment. Portals provide a way for enterprises to provide a consistent look and feel with access control and procedures for multiple applications and databases, which otherwise would have been different entities altogether. Examples of public Web portals are Yahoo! and iGoogle. Different types of Web portals are personal portals, government Web portals, news portals and corporate Web portals.

Web Log: Web logs are a large database stored in the Web servers that contain details of the transactions of Web users. There are many fields in the Web log database, which conform to either Common Log Format (CLF) or the Extended Common Log Format (ECLF). The fields specified by this format are IP address of the destination page, Destination URL, Code which denotes the packet number, Protocol which specifies the type of Network protocol used (e.g. TCP), Method which specifies the type of method HTTP method used (e.g. GET, POST), Type which specifies the file type, Date which gives the date and time when the page was accessed, Referrer which gives the IP address of the client which requested the page, and finally the Size of the page in bytes. Web logs are very useful for a variety of applications such as predicting Web users' behaviours and in this chapter we have focussed on using Web logs to derive the visited URLs to map them efficiently onto their appropriate categories using LSW.

Data Preprocessing: This process removes irrelevant and redundant information or noisy and unreliable data such as images, media and scripts. Data preparation and filtering steps can take a considerable amount of processing time. Here, we first extract the Destination URLs from Web log and crawl the entire page from the Web server. We extract only text and eliminate Images, Scripts, from the crawled Web pages. From the text obtained we remove stop words, BVerbs and also apply stemming for words.

Web Directory: A Web directory or link directory is a directory on the WWW that specializes in linking to other Web sites and categorizing those links. A Web directory is a human-edited search engine and does not display lists of Web pages based on keywords; instead it lists websites by category and subcategory. Most Web directory entries are also not found by Web crawlers but by humans. The categorization is usually based on the whole website rather than one page or a set of keywords, and sites are often limited to inclusion in only a few categories. Web directories often allow site owners to submit their site for inclusion, and have editors review submissions for fitness. A human-edited directory is created and maintained by editors, who add links based on the policies particular to that directory. Submission of websites to Web directories is considered a common search engine optimization technique to get backlinks for the submitted Web site.

Categories: It is a cluster of related pages. These clusters are extracted from the Web directory for mapping of browsed pages of users to them.

Hash Engine: This calculates the hash values for the given Web pages of user interest and pages in categories. The pages are compressed and stored in the form of strings, where one string represents one page.

MPC-LSW: From the obtained strings of used Web pages and pages in categories, Levenshtein Similarity is applied between them and mapping of pages to categories is performed.

5.4 Mathematical Model and Algorithm

Consider a set of Web pages $w = \{p_1, p_2, \ldots, p_n\}$ that are downloaded using the visited URLs of a Web log, where p_i is the browsed Web pages, $\forall i$, $1 \leq i \leq n$ and set of m categories C=$\{c_1, c_2, \ldots, c_m\}$ where each category c_i is the categories in the Web directory $\forall i$, $1 \leq i \leq m$, that may contain K pages $\{p_{i1}, p_{i2}, p_{i3}, \ldots, p_{ik}\}$ where $1 \leq k \leq l$, where l is maximum number of pages in c_i. for example, if $C = \{c_1, c_2, c_3, c_4\}$ and if c_3 has the highest number of pages, then $l = |c3|$. Similarity between page $p \in W$ and category $c_j \in C$ is calculated using Hashing, Levenshtein Distance (LD) and Levenshtein Similarity weight (LSW). Each of these processes is discussed in the following sections.

5.4.1 Hashing

An Hash function $H(t)$ for compressing term t is given by [9] and represented as

$$H(t) = (a_1 x^{k-1} + a_2 x^{k-2} + \cdots + a_k x^0) mod p \quad (5.1)$$

Where $A=\{a_1, a_2, a_3, \ldots, a_k\}$ are the co-efficient of polynomial which are ASCII equivalents of alphabets, x is an integer value, which permits exponentiation and term-wise explanation and p is the number of letters in an alphabet, for English $p = 26$.

Compressing a line L_i consisting of m terms, $H(t)$ is computed for each term $t \in L_i$, that generates a set of hash values represented as

$$H(L_i) = \{H(t_i) : 1 \leq i \leq m\}. \quad (5.2)$$

For each $H(t_i) \in H(L_i)$, we calculate the ASCII equivalents to obtain the represented as

$$H(L_i^{'}) = \{H(a_i^{'}) : 1 \leq i \leq m\}. \quad (5.3)$$

Finally, we concatenate the elements in $H(L_i)$ to get a compressed representation of line L_i to obtain string S_i.

5.4.2 Levenshtein Distance (LD)

The Levenshtein Distance [31] is the number of operations (insertions, deletions and substitutions) required to transform a string S into another string P. LD can be computed across a hierarchy of letters, words, phrases, sentences, paragraphs, or even pages. LD between two strings S and P corresponding to two Web pages is given by the recurrence equation

$$d[i, j] = \begin{cases} i, & \text{if j = 0} \\ j, & \text{if i = 0.} \\ d[i-1, j-1], & \text{if s[i] == t[j]} \\ min\{d[i-1, j], d[i, j-1], d[i-1, j-1]\} + 1 & \text{otherwise} \end{cases} \quad (5.4)$$

Where $d[i, j]$ is the number of operations required to convert a string S with first i characters and string P with j characters, where i is an index for the string S and j is that of P.

5.4.3 Levenshtein Similarity Weight (LSW)

LSW between line L_i and L_j is given by

$$LSW(L_i, L_j) = 1 - \frac{LD(L_i, L_j)}{maxlength(L_i, L_j)}. \tag{5.5}$$

LSW(L_i, L_j) is 0 when there is absolutely no match at all between L_i and L_j, 1 when L_i and L_j are exactly the same. LSW(L_i, L_j) always lies between 0 and 1, in all other cases. For example, LSW (PQR, LMN) = 0, LSW (PQR, PQR) = 1 and LSW (PQRS, QRS) = 1 − (1/4) = 0.75.

5.4.4 Similarity Between Web Page and Category in Web Directory

Consider a page P and for each line $l \in P$ find the hash value $H(t_i)$ for each term $t_i \in l$ using Eq. (5.1). By combining all hash values of terms present in line l, we get compressed representation of l, i.e. $H(l)$ using Eq. (5.2) and its ASCII equivalent $H(l)$ is calculated using Eq. (5.3) to get a single term corresponding to line l. Similarly, this procedure is repeated for all lines in the Web page leading to a set of terms which are concatenated to one string S, which in turn helps in further processing. This procedure is repeated for all the browsed Web pages and for all the pages in categories.

Example 5.1 Consider a Web page with links, images, text in tables and text with lot of unwanted information. From this, only text portion of the page is extracted using data extraction process, represented in first column of Table 5.1. For the extracted text, BVerb removal and stemming process are applied thus resulting in second column. Compressed representation of second column data is as shown in third column after applying Eqs. (5.1), (5.2) and (5.3).

Example 5.2 Consider a category C with pages 1.html and 2.html, text is extracted and cleaned for which hashing function is applied as shown in columns 2, 3 and 4, respectively, of Table 5.2.

Using this procedure, a page P is translated into one term t and for all pages $p_1, p_2, \ldots, p_k$ in category c_i translated into set of terms $ct_1, ct_2, \ldots, ct_k$. Similarity between P and c_i is calculated using Eq. (5.6).

$$sim(P, c_i) = \sum_{i=1}^{k} LSW(t, ct_i). \tag{5.6}$$

Table 5.1 Browsed page and its hash value

Uncleaned Web Page	Cleaned page	Hashed value of page
In just 4 years of captaincy, Mahendra Singh Dhoni has emerged as one has as one of India's most successful captain this cricket crazy country has seen. The Captain Cool has led the team to many first under his captaincy. Dhoni led Team India to No. 1 in the ICC Test Rankings for the first time in December 2009 and also closed the gap with Australia in one-dayers after winning the coveted 2011 Cricket World Cup.	just 4 year captaincy, mahendra singh dhoni emerge india's success captain cricket crazy country seen. captain cool led team captaincy. Dhoni led team india 1 icc test rank time December 2009 close gap australia -dayer win covet 2011 cricket world cup.	ayaqkhydzbhyujvqyggx mdgxruxpopdkrzqjoqmlucw

5.4.5 *Mapping of Pages onto Categories*

Mapping of browsed pages to the category in the Web directory is achieved by finding similarity between page $P \in W$ and all categories in C using Eq. (5.6). Next the page P is mapped to the correct category C_p, that has maximum similarity which is computed using Eq. (5.7).

$$c_p = argmax\{sim(P, c_i) : \forall c_i \in C\}. \tag{5.7}$$

Example 5.3 Consider a browsed Web page P and three categories c_1 with 3 pages p_1, p_2, p_3, category c_2 with 2 Web pages p_4, p_5 and category c_3 with p_6, p_7, p_8, p_9. LSW between browsed page P and pages in category c_1 are 0.18667, 0.16667, 0.14667 and average similarity c_1 and page P is 0.16667. LSW between browed page p and pages in category c_2 are 0.16667, 0.17333 and average similarity of c_2 and page p is 0.16995. LSW between browed page p and pages in c_3 are 0.17333, 0.18667, 0.19334, 0.17333 and average similarity of c_3 and page p is 0.18675. Considering the average similarities of all categories, it is found that maximum similarity is between page P and category c_3. Hence, page P is mapped to category c_3.

5.4.6 *Algorithm MPC-LSW*

Algorithm for mapping of Page onto Category using Levenshtein Similarity Weight (MPC-LSW) is shown in an Algorithm 5.1. It accepts Web page P and set of categories C as inputs and returns a pair of page and category as output which represents

Table 5.2 Category pages and their hash values

Page No.	Uncleaned page	Cleaned page	Hashed value of page
1.html	Known as 'the Wall' due to his ability to bat for long, bat long, Rahul Dravid has been the batting mainstay of the Indian Test team since he first arrived at the international scene in 1996. Dravid is regarded as one of the most technically sound batsmen of his time. Dravid holds the record of being the only batsman to score a century in all ten Test playing nations	Known 'wall' due abil ahul dravid bat mainstai indian test team arriv intern scene 1996. dravid regard technic sound batsmen time. dravid hold record batsman score centuri test plai nation	Mgauqonei Gchyzrxchsiu Eyphcsuqo jhwasyu
2.html	MUMBAI:Master blaster Sachin Tendulkar has finally responded to the BMC's invite. After he notched his 100th century, the BMC on April 3 had passed a congratulatory motion in the civic House, esting the cricketer to make himself available for a felicitation programme. Nearly three months on, Tendulkar finally responded but much to the dismay of civic officials, he has not made any commitment with regard to the felicitation plans	Mumbai: master blaster sachin tendulkar final respond bmc's invit. notch 100th centuri bmc April 3 pass congratulatory motion civic house, request cricket avail felicit programm. nearly month, tendulkar final respond dismai civic official, commit regard felicit plan	Ennheaeeea booxyediesp gedegqwzfops

highest similarity between the page and category in the pair. Firstly, page P is compressed into string HP using function $hash(p)$. Next, all the pages p_1, p_2, p_3,..., p_k in the category c_i, are also compressed into string HP_1, HP_2, HP_3,..., HP_k respectively. Finally, average of similarities between HP and $(HP_1, HP_2, HP_3, \ldots, HP_k)$ is found. In the same way, similarity between page p and all categories in c is calculated and stored in array $sim[1..m]$, where $sim[i]$ stores the similarity value between p and c_i. Finally, we examine the maximum similarity using function $Max(sim[1..m])$ and based on the maximum value obtained, we assign the page P to the appropriate category.

Algorithm 5.1: MPC-LSW Algorithm.

Input : Browsed page $P \in W$, Set of categories $C = \{c_1, c_2, \ldots, c_m\}$. where c_i has K pages.

Output : Page category pair (P, c_i), where P is more affine to category c_i

```
begin
    //***********Computes similarity between P and ci*************
    String HP=hash(P);
    for each category ci ∈ C do
        Sim[i]–0;
        for each Pj ∈ ci do
            HPj=hash(Pj)
            Sim[i]+=Similarity(HP,HPj);
        Sim[i]/=2;
    maxVal=Max(Sim);
    return (P,maxVal);
//**********Computes maximum of Similarities***************
Function Max(Sim[1..m])
begin
    Max=1;
    for (i=1 to m-1) do
        if (sim[i]<sim[i+1]) then
            Max=i+1;
    return Max;
//***********Computes hash value************************
Function hash (P)
begin
    String S = ;
    for each line l ∈ P do
        si=;
        for each term t ∈ l do
            H[tj]=Compute Hash value H(t);
            H[tj]=Invert H[tj] to ASCII equivalent.
            si = Concatenate(si, H(tj));
        S = Concatenate(S, si);
    Return S;
```

5.5 Experiments

The MPC-LSW algorithm is implemented using Perl and Shell scripting language. Ubuntu Operating System with built-in Perl in a system running on Intel core i3 with 3 GB RAM and 320 GB HDD. The dataset consists of categories like Sports, Entertainment, Politics, etc., containing sets of related pages and browsed pages extracted from the Web log. The browsed pages are assumed to get mapped onto any one of these existing categories. The synthetic dataset consists of 54 categories with an average of 22 pages in each category onto which 280 browsed Web pages are to

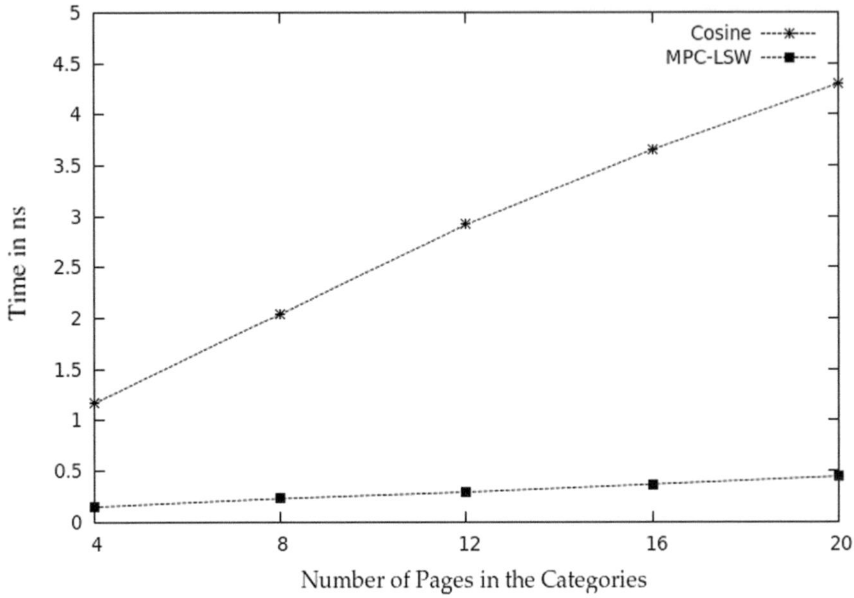

Fig. 5.2 Execution time

be mapped. The importance of the algorithm is to find the similarity between each browsed Web page and all the categories. Based on the maximum similarity, mapping of the particular page to the appropriate category is performed. The algorithm MPC-LSW has been executed for different numbers of pages and has been compared with TF-IDF-based Cosine Similarity algorithm. The following sections discuss the parameters used for evaluating the performance of MPC-LSW.

5.5.1 Execution Time

The graph is plotted for the number of pages in categories versus time in *ns* which gives execution time between Cosine similarity and MPC-LSW for mapping of browsed pages to topic directory categories as shown in Fig. 5.2. From the graph, it can be observed that Cosine Similarity takes more time than MPC-LSW since the former's time complexity is $O(mkn)$ and that of the latter is $O(mk)$. Hence, the gap between the line graphs of MPC-LSW and Cosine Similarity is increasing with the increase of categories indicating that Levenshtein similarity based algorithm runs faster than the TF-IDF based ones.

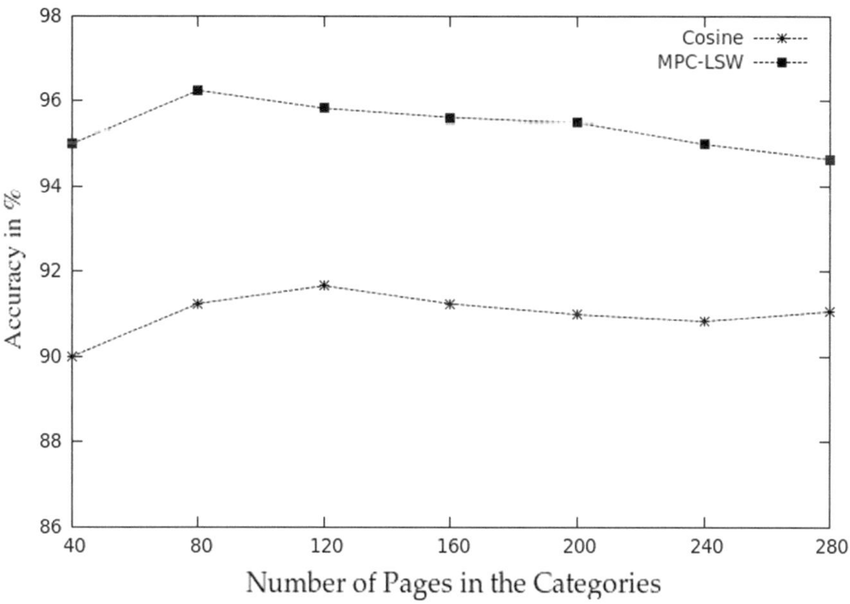

Fig. 5.3 Accuracy comparison

5.5.2 *Accuracy*

Accuracy is the measure that gives us the percentage of pages that are correctly classified by the algorithm. The Web pages are classified by mapping them onto existing categories or by creating new categories. Accuracy is computed using

$$Accuracy = \left(1 - \frac{Pages\ classified\ wrongly}{Total\ no.\ of pages}\right) * 100. \tag{5.8}$$

Figure 5.3 clearly shows that the accuracy of MPC-LSW is 5–0% greater than that of Cosine Similarity. This is due to the fact that the faster execution time along with the string-representation of the Web pages results in a more accurate and reliable method for mapping the Web pages to their appropriate categories.

5.5.3 *Precision*

Precision is the measure that gives us the percentage of correctly mapped pages to the total number of mapped pages. Precision is computed using

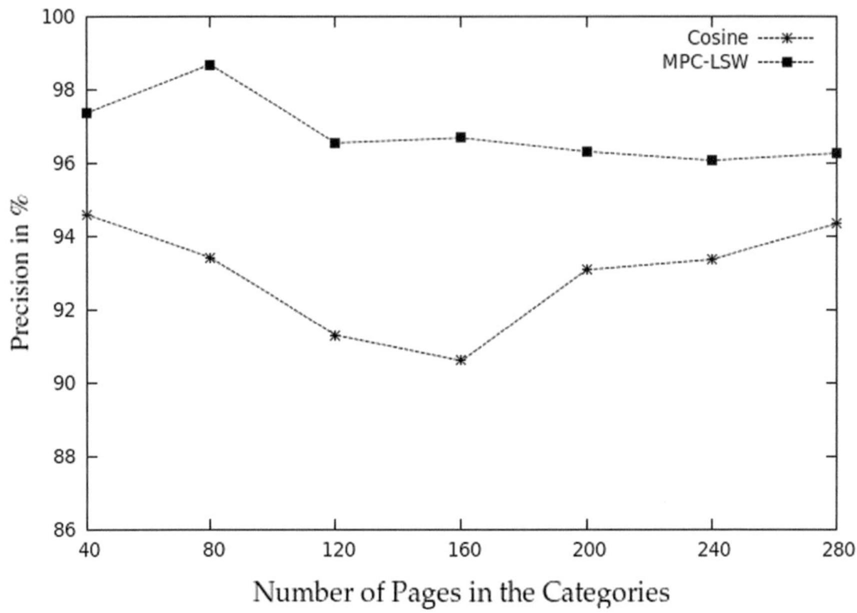

Fig. 5.4 Precision comparison

$$Precision = \frac{Pages\ mapped\ correctly}{Total\ no.\ of\ mapped\ pages} * 100. \tag{5.9}$$

Figure 5.4 shows that the Precision of MPC-LSW is 2–10% greater than that of Cosine Similarity. Precision fluctuates significantly as it is possible for the newly created categories for those pages which do not come under the existing categories to accommodate newer pages that satisfy the criteria to be placed under them. The higher Precision rate of MPC-LSW illustrates the reliability of our algorithm even as it takes lesser time to execute than Cosine Similarity.

5.5.4 Recall

Recall is the measure that gives us the percentage of the correctly mapped pages to the total number of Web pages. Recall is computed using

$$Recall = \frac{Pages\ mapped\ correctly}{Total\ no.\ of\ pages} * 100. \tag{5.10}$$

Figure 5.5 illustrates that the Recall of MPC-LSW is 3–10% higher than that of Cosine Similarity. Of all the given Web pages, knowing how many pages can

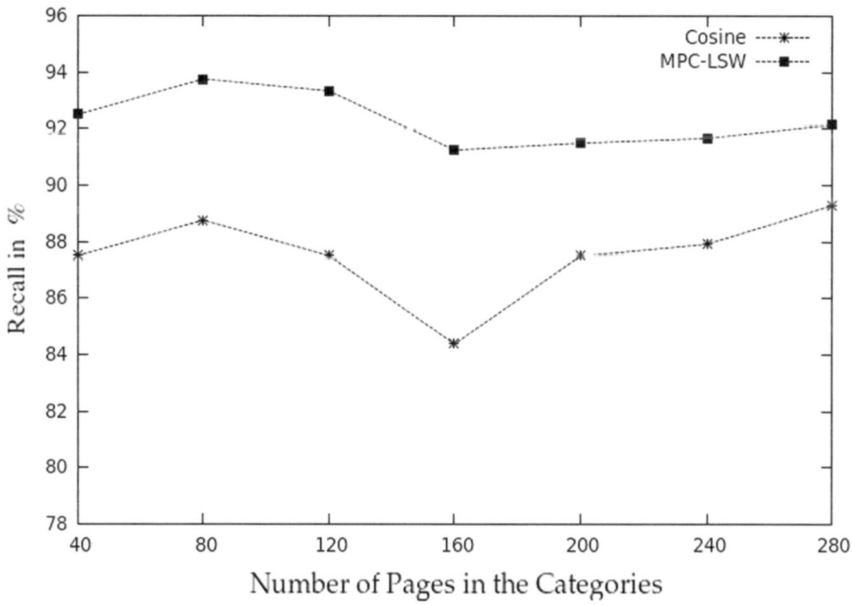

Fig. 5.5 Recall comparison

be mapped correctly is beneficial for the effective and accurate construction and mapping of topic directories. By choosing methods with high Recall such as MPC-LSW, it results in restricting the number of wrongly mapped pages as small as possible.

5.5.5 *F-Score*

The F-score is an optimized measure that combines both Precision and Recall into a single parameter without losing either of their importance. It is a standard measure used universally to determine the efficiency of a mapping function. F-score is computed as

$$F\text{-}score = 2 * \frac{Precision * Recall}{Precision + Recall}. \tag{5.11}$$

Figure 5.6 illustrates the difference of 2–10% in the F-scores of MPC-LSW and Cosine Similarity because the former performs page-level comparison whereas the latter performs term-level comparison. Since the F-scores efficiently summarize the effects of Precision and Recall, this concludes that MPC-LSW is more efficient.

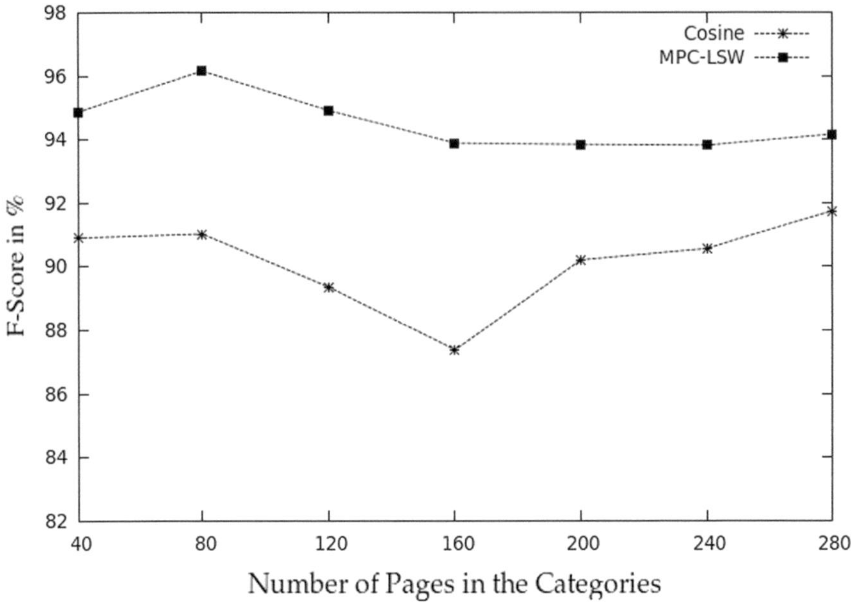

Fig. 5.6 F-score comparison

5.6 Summary

In this chapter, we propose a new algorithm called MPC-LSW for mapping of the browsed pages onto topic directories based on Levenshtein similarity. LSW is calculated after compressing contents of browsed pages and category pages using hash function and LD. Then, the browsed page is mapped onto the category based on the maximum LSW. The experimental results obtained support the fact that the proposed algorithm runs faster than the algorithms based on TF-IDF. Accuracy and F-score of MPC-LSW are also moderately higher than that of Cosine Similarity. This algorithm can be used as time-efficient similarity measure for construction of community Web directories, classification and clustering of Web pages and further for personalization of Web directories.

References

1. D. Pierrakos, G. Paliouras, Personalizing Web Directories with the Aid of Web Usage Data. IEEE Trans. Knowl. Data Eng. **22**(9), 1331–1344 (2010)
2. H. Hyyro, A bit-vector algorithm for computing Levenshtein and Damerau edit distances. Nord. J. Comput. **10**(1), 29–39 (2002)
3. S. Burkhardt, J. Karkkainen, One gapped q-gram filters for Levenshtein distance, in *CPM*. Springer LNCS, vol. 2373 (2002) pp. 225–234

4. S. Rane, W. Sun, Privacy preserving string comparisons based on leveinstein distance, in *Proceedings of IEEE International Workshop on Information Forensics and Security* (2010), pp. 1–6
5. M. Bilenko, R. Mooney, W. Cohen, P. Ravikumar, S. Fienberg, Adaptive name matching in information integration. IEEE J. Intell. Syst. **18**(5), 16–23 (2003)
6. S. Mihov, S. Koeva, C. Ringlstetter, K.U. Sculz, C. Strohmailer, Precise and efficient text correction using levenshtein automata, dynamic web directories and optimized correction models
7. J.G. Fiscus, J. Ajot, N. Radde, C. Laprun, Multiple dimension levenshtein edit distance calculations for evaluating automatic speech recognition system during simultaneous speech
8. K Sankar, L Sobha, An approach to text summarization, in *Proceedings of CLIAWS3, 3rd International Cross Lingual Information Access Workshop* (2009), pp. 53–60
9. R. Mishra, P. Kumar, Clustering web logs using upper approximation with different similarity measures. Int. J. Mach. Learn. Comput. **2**(3), 219–221 (2012)
10. A. De Lucia, M. Risi, G. Scanniello, G. Tortora, Towards automatic clustering of similar pages in web applications, in *Proceedings of 11th IEEE International Symposium on Web Systems Evolution (WSE)* (2009), pp. 99–108
11. S.K. Fathy, Automatic classification of web pages. IJICIS **4**(1), 26–34 (2004)
12. N. Soonthornphisaj, B. Kijsirikul, in *Proceedings of World Academy of Science, Engineering and Technology*, vol. 1 (2005) pp. 120–123
13. D. Shen, Z. Chen, Q. Yang, H.-J. Zeng, B. Zhang, Y. Lu, W.-Y. Ma, Web page Classification through Summarization, SIGIR (2004)
14. Z. Lin, K. Deng, Y. Hong, Research of web pages categorization, in *Proceedings of IEEE International Conference on Granular Computing* (2007) pp. 691–694
15. D. Riboni, *Feature Selection for Web Page Classification*
16. S. Lakshminarayana, Categorization of web pages-performance enhancement to search engine. Elseveir J. Knowl. Based Syst. **22**, 100–104 (2009)
17. M. Javid, A.H. Sabery, A. Khanteymoory, Ant colony algorithm for web page classication, in *IEEE International Symposium on Information Technology*, vol. 3 (2008) pp. 1–8
18. S. Bo, S. Qiurui, C. Zhang, F. Zengmei, A study on automatic web pages categorization, in *Proceedings of IEEE International Advance Computing Conference (IACC-09)* (2009) pp. 1423–1427
19. S.-M. Kim, P. Pantel, L. Duan, S. Gaffney, Improving web page classification by label-propagation over click graphs, in *Proceedings of the 18th ACM Conference on Information and Knowledge Management (CIKM '09)* (2009) pp. 1077–1086
20. A.P. Asirvatham, K.K. Ravi, *Web Page Categorization based on Document Structure* (2008) http://citeseer.ist.psu.edu/710946.html
21. J.M. Pierre, On the automated classification of web sites. Linkop. Electron. Artic. Comput. Inf. Sci. **6**, 1–12 (2001)
22. H. Yu, K.C.-C. Chang, J. Han, Heterogeneous learner for web page classification, in *Proceedings of the IEEE International Conference on Data Mining (ICDM '02)* (2002) pp. 538–545
23. H. Yu, K.C.-C. Chang, J. Han, PEBL: Web page classification without negative examples. IEEE Trans. Knowl. Data Eng. **16**(1), 1–12 (2004)
24. R. Prabowo, M. Jackson, P. Burden, H.-D. Knoell, Ontology-based automatic classification for the web pages: design, implementation and evaluation, in *Proceedings of the 3rd IEEE International Conference on Web Information Systems Engineering* (2002) pp. 182–191
25. M. Tsukada, T. Washio, H. Motoda, Automatic web-page classification by using machine learning methods, in *Proceedings of the First ACM Asia-Pacific Conference on Web Intelligence: Research and Development* (2001) pp. 303–313
26. A.S. Patil, B.V. Pawar, Automated classification of web sites using nave bayesian algorithm, *IMECS*, vol. 1 (2012) pp. 466–470
27. X. Peng, B. Choi, Automatic web page classification in a dynamic and hierarchical way, in *IEEE International Conference on Data Mining* (2002) pp. 386–393
28. M. Chau, P.Y.K. Chau, P.J. Hu, Incorporating hyperlink analysis in web page clustering, in *Proceedings of the Sixth Workshop on E-Business* (2007) pp. 844–849

29. C. Lin, Y. Yu, J. Han, B. Liu, Hierarchical web-page clustering via in-page and cross-page link structures, in *4th International Conference on Information Processing (ICIP)* (2012) pp. 120–130
30. P. Shi, An efficient approach for clustering web access patterns from web logs. Int. J. Adv. Sci. Technol. **5**, 1–13 (2009)
31. http://en.wikipedia.org/wiki/Levenshtein_distance

Chapter 6
Related Search Recommendation with User Feedback Session

K. R. Venugopal and Sejal Santosh Nimbhorkar

Abstract Keyword-based search is an extensively used method to discover knowledge on the Web. Generally, Web users are unable to arrange and define input queries relevant to their search because of adequate knowledge about the domain. Therefore, the input queries are normally ambiguous and short. Query suggestion is a method to recommend queries related to the user input query that helps them to locate their required information more precisely. It helps the search engine to provide relevant answers and meet users needs. Usually, users query keywords are ambiguous, therefore it is not good to use users query keyword in suggestion. In this chapter, Related Search Recommendation (RSR) framework is presented that determines keywords presented in un-clicked and clicked documents in the feedback session. Feedback sessions are used to retrieve users need in terms of Pseudo documents. Semantic similarity is computed between the terms in the Pseudo document. The semantic terms are used for suggestions. The presented method provides semantically related search queries for the user input query. Results show that the RSR method outperforms Rochios model and Snippet Click Model.

6.1 Introduction

Web data keeps expanding and is available in various data forms because of the rapid growth of online advertising, publishing, e-commerce and entertainment. Although Web search technology provides efficient and effective information access to users, it is still a difficult task to search useful knowledge about user needs from their search queries. Therefore, query suggestion is an important and an essential feature of commercial Web search engines. The users can directly use query suggestions results for the future new search.

K. R. Venugopal (✉)
Bangalore University, Jnana Bharathi, Bengaluru 560056, India
e-mail: venugopalkr@gmail.com

S. Santosh Nimbhorkar
BNM Institute of Technology, Banashankari, Bengaluru 560070, India
e-mail: sej_nim@yahoo.co.in

K. R. Venugopal et al., *Web Recommendations Systems*,
https://doi.org/10.1007/978-981-15-2513-1_6

Query suggestion is an efficient way to enhance keyword-based search, which is extensively useful for Web search systems. Users need to modify queries so often because queries are often informational. Users may seek discrete information on a distinct subject, hence may check out various query terms. Users may not have sufficient knowledge on a topic, therefore adequate terms are not known to retrieve the required information.

In Kato et al. [1], query recommendations are frequently used when (1) a initial query is an exceptional query, (2) single-term query is used as input query, (3) explicit queries are suggested, (4) suggestions are provided based on modification of input query and (5) various URLs has been clicked by users on the resulting search page.

Query suggestions provided to the user efficiently can reduce the complexity of the search and help them to locate the required information more precisely. This method is extensively accepted by product, music, video search, retrieval of medical information and patent search information. Query suggestions techniques are implemented by commercial search such as *Searches related* in Google, *Search Assist* in Yahoo! and *Related Searches* in Bing Search.

Through query suggestion, search engines have succeeded in obtaining Web information for users, but the keyword-based search is not able to organize and formulate input queries. Silverstein et al. [2] derived that users' input query's average length is 2.35 terms (AltaVista search engines query log). This shows that most of the user queries are short. A short query cannot describe the information needed of user search and sometimes ambiguous in meaning expression. Because of insufficient knowledge about domain, users find it difficult to organize and define appropriate input queries. Then the user has to rephrase the query words or query frequently, which affects the search performance.

In ([3–8]), the authors have focused on query suggestions by considering users' previous query and click behaviour. There are two major issues with query-URLs recommendations: (i) the common clicks on URLs are limited for various queries (ii) though users may click the same URLs for two different queries, they may be irrelevant as that Web documents may have different contents [9]. It is necessary to generate useful suggestions by solving these problems. It is required to discover users' information needs to organize queries with a precise meaning. Users' search log provides information needs from users' click behaviour. If a certain retrieved result is clicked by the user, we cannot conclude that the clicked result is completely relevant to the user query since he has not seen the full document. But the brief description of the document, i.e. snippet is shown to the user and is read by the user if he decides to click that document. It can be considered that snippet reflects the user's information need.

Lu et al. [10] have designed a method to determine the goal of user search for a query. These search goals are obtained by clustering the proposed feedback sessions. The clicked and un-clicked documents with the last clicked document represents feedback session in a user search log. Pseudo documents are obtained by mapping feedback sessions to reflect the information needs of the users effectively. Pseudo documents are clustered using k-mean algorithm to derive search goals of users.

In this chapter, Related Search Recommendation (RSR) framework is proposed to recommend related queries for user input query. This framework uses user feedback from click through log of search engine. User click through log is converted into feedback session with clicked and un-clicked URLs and it ends with last clicked URL in a session. Each clicked and un-clicked URLs of feedback session is converted into enriched documents by calculating term frequency–inverse document frequency for each term present in the title and snippet of that URL. Pseudo documents are generated by merging all the enriched documents of a feedback session. Finally, the optimized pseudo document is generated by combining all the pseudo documents for a given input query, which reflects the user's information need. Recommendations are generated and ranked by combining query and terms for all the methods.

6.2 Related Works

Mostly, users access Web pages by querying through search engines by which the performance of search engines is affected. In this chapter, we are recommending related search queries with the user feedback session. In this session, clicked and un-clicked document's snippets are used to formulate related search queries. We need to calculate the similarity between different words that exist in snippets to obtain the desired results. We have reviewed several papers related to measuring the similarity between words and different techniques used for query recommendations using snippets in this section.

6.2.1 Measuring Similarity Between Two Words

Miao et al. [11] have developed a query expansion method based on Rocchio's model. In this model, proximity information is modelled by proposed Proximity-based Term Frequency *ptf* in the pseudo relevant documents. Expansion terms and their proximity relation with query terms is modelled by *ptf*. Window-based, kernel-based and Hyperspace Analogue to Language (HAL) methods are proposed as proximity measures for evaluating the relationship between query terms and expansion terms. This model achieves better performance over position relevance model and classic Rocchio's model.

Hamai et al. [12] have discussed a transformation function to measure semantic similarity between two given words. This approach uses page counts of documents title to measure similarity. This approach outperforms similarity measures defined over snippets.

Bollegala et al. [13] have presented an approach to calculate semantic similarity between words. Text snippets are used to obtain Lexico-syntactic patterns from a Web search engine. Support vector machine is used to integrate page count based similarity score and lexico-syntactic patterns to generate semantic similarity measure.

This method performs better than Information content measures and Edge counting WordNet-based methods.

Li et al. [14] have presented an approach to calculate the semantic similarity between terms and multiword statement. A large Web corpus is used to form an *isA* semantic network to provide contexts for the terms. The meaning of input terms is formulated by K-medoids clustering algorithm and similarity is computed with *max–max* similarity function. This algorithm outperforms multiword expression pairs and Pearson correlation coefficient on word pairs.

Bollegala et al. [15, 16] have developed a relational model to calculate the semantic similarity between two words. Snippets of Web pages are used to obtain lexical patterns. Semantically related patterns are identified by extracted clusters from sequential pattern clustering algorithm. Mahalanobis distance is used to calculate semantic similarity between two words. This method outperforms all WordNet-based approaches ([17–22]).

6.2.2 Query Recommendation Techniques

Song et al. [23] have designed query suggestion method by using users' feedbacks in the query logs. Query-URL bipartite graph is constructed for click and un-click information. Random Walk with Restart (RWR) technique is applied to both the graphs. The category of URLs is used to construct correlation matrix for URLs. Optimal query correlation matrix is constructed by combining two query correlation matrices, which is used for query suggestion. This framework gives better performance than pseudo-relevance feedback models ([24–26]) and random walk models.

Kharitonov et al. [27] have focused on contextualization framework for diversifying query suggestion. This framework utilizes the user's history query, the previously clicked and skipped documents and examines query suggestions. Mean Reciprocal Rank (MRR) is used as a performance evaluation metric. This framework is compared with non-diversified ranking with the previous query, ranking with the previous query as a context and clicks and skips as context.

Ozetem et al. [28] have developed an approach to learn the probability with machine learning that a user may find a relevant follow-up query after executing the input query. To measure the relevance of follow-up query, probabilistic utility function is used which relies on the query co-occurrence. To capture the semantic similarity of the suggestions, lexical and result set based characteristics are developed. Gradient Boosted Decision Tree (GBDT) regression is performed to rank the suggestions for input query and remove the irrelevant. This approach shows significant improvement over Mutual Information (MI) method.

Broccolo et al. [29] have investigated a query suggestion algorithm that can cover long tail queries. This algorithm uses search shortcuts model to process a full text query, which is indexed in user sessions recorded in a query log. This algorithm outperforms Query Flow Graph (QFG) and Cover Graph (CG) by providing the most relevant query suggestions.

Zhang et al. [30] have developed an approach for query suggestion based on query search. This approach constructs an ordered set of search terms drawn from documents to create candidate query suggestions. It builds query suggestions separately for each potentially relevant document. This approach provides more relevant query suggestions for short queries as well as long queries.

Gomex et al. [31] have designed a novel technique to visualize the collection of textual snippets returned from a Web query. This technique constructs intuitive and meaningful layouts that optimize the placement of snippets by employing an energy function. This function considers both overlapping removal and preservation of neighbourhood structures. This technique is compared with VPSC, PRISM, Voronoi based and RWordle-C by using Euclidean distance, layout similarity and neighbourhood preservation metrics.

Phan et al. [32] have introduced a method to process sparse and short documents by hidden-topic-based framework on the Web. This framework solves data sparseness and synonyms/homonyms problems of documents. Common hidden topics are determined from datasets to make documents short, less sparse and more topic oriented. This framework is evaluated for online advertising applications on Web search domain matching/ranking and classification. Precision and recall are used to evaluate hidden topics which are used in the improvement of ranking and matching performance.

He et al. [33] have presented a novel sequential query prediction approach for understanding users' search intent and recommending queries. A sequential probabilistic model called Mixture Variable Memory Markov Model is developed for online query recommendation. Experiments results show that ordered queries within the same session are highly correlated and should be utilized to understand the user information needs. Coverage and accuracy are used as performance evaluation metrics.

Jiang et al. [34] have presented a query recommendation method based on Query Hashing (QH). QH generates many similar and dissimilar query-pairs as prior knowledge from query sessions. Then QH learns a transformation from the prior knowledge such that after transformation of similar queries tend to have similar hash values. In the recommendation stage, queries that have similar hash values to the given query are ranked and *top K* queries are displayed as the recommendation result. QH model is compared with hashing-based methods, SimHash, Kernelized Locality Sensitive Hashing and Inverted list. This method achieves the best results in terms of efficiency and recommendation performance.

Li et al. [35] have proposed a query suggestion approach. In the learning step, a generative probabilistic model is obtained by learning external knowledge gained from the Web dataset for Web queries. Latent semantic topic model is used to organize the co-occurrence of the Web queries. Posterior distribution of hidden topics is obtained for each candidate query with this model. The topic distribution is acquired in online query suggestion step for an given input query. The candidate queries and input query similarity is computed by using their corresponding topic distribution. Finally, suggestions are provided by listing candidate queries based on similarity score. Precision and Mean Average Precision (MAP) is used as evaluation metrics.

Table 6.1 Related work comparison

Author	Concept	Advantages	Disadvantages
Li et al. [35]	Suggest topically related Web queries using hidden topic model	Provides better query suggestions than URL model and comparable results with term feature model	Training dataset need to be generated to find topic of Web queries from external data source
Zhang et al. [30]	Provide improved query suggestion by query search	Provides more relevant query suggestion for short queries as well as long queries compared to suggestion by query search	User feedback is not considered
Miao et al. [11]	Query expansion based on proximity based Rocchio's model	This model achieves better performance over position relevance model	The exact relationship between the window size factor and information of collection is not fixed
Lu et al. [10]	Inferring user search goals with feedback sessions	User search goals can be utilized in query recommendations	Finds personalized search goals
Liu et al. [36]	Provide query recommendation based on snippet click model	Provides more effective recommendations than Biadu and Sogou search engines	Only click information is used to create model
Rocchio [37]	Query expansion with user feedback	Considers user feedback and generates relevant terms for query expansion	Fails to classify multimodal classes and relationship
RSR	Recommending related search with user feedback session and semantic similarity between words	Provides semantically related search to inputs and this approach can be extended to generate multiple related words	

This approach gives better query suggestions than URL model and comparable results with the term feature model.

Liu et al. [36] have proposed a snippet click model for query recommendation. This model determines the information need of users from search logs. The clicked snippets are used to represent the information need of the users and with this judgement snippet click models are constructed. Click through rate and click amount are used as metrics to evaluate the performance of the algorithm. The proposed algorithm is providing more efficient recommendation than Biadu and Sogou search engines.

Table 6.1 shows comparison of related works.

6.3 Related Search Recommendation Framework and RSR Algorithm

6.3.1 Problem Definition

Given a user input query q and user click through log lg from the Web search engine S, our objective is to recommend expanded queries q_e. It is assumed that the user is online while entering input query and considers only top-50 retrieved search results.

6.3.2 Co-occurrence Measures to Compute Semantic Similarity

Co-occurrence measures Dice, Jaccard, Pointwise Mutual Information (PMI) and Overlap are explained to calculate semantic similarity. The notation *P(Q)* is used to represent the page counts for the query Q in the search engine. The *WebJaccard* between terms T_1 and T_2, (i) *WebJaccard*(T_1, T_2) is defined as

$$WebJaccard(T_1, T_2) = \frac{P(T_1 \cap T_2)}{P(T_1) + P(T_2) - P(T_1 \cap T_2)} \tag{6.1}$$

Here, $P(T_1 \cap T_2)$ denotes the co-occurrence of terms T_1 and T_2.
(ii) *WebDice*(T_1, T_2)*WebDice* is defined as

$$WebDice(T_1, T_2) = \frac{2P(T_1 \cap T_2)}{P(T_1) + P(T_2)} \tag{6.2}$$

WebOverlap(T_1, T_2) is a natural modification to the Simpson coefficient. (iii) *WebOverlap*(T_1, T_2) is defined as

$$WebOverlap(T_1, T_2) = \frac{P(T_1 \cap T_2)}{min(P(T_1), P(T_2))} \tag{6.3}$$

Pointwise Mutual Information (PMI) is a measure of association used in statistics and information theory. It reflects the dependencies of two probabilistic events. (iv) *WebPMI* is defined as a modification of pointwise mutual information using page counts as

$$WebPMI(T_1, T_2) = \log_2\left(\frac{\frac{P(T_1 \cap T_2)}{N}}{\frac{P(T_1)}{N}\frac{P(T_2)}{N}}\right) \tag{6.4}$$

6.3.3 WordNet-Based Semantic Similarity

WordNet based measures are discussed to calculate semantic similarity. WordNet [38] developed by Princeton University is a lexical database in English. It is well suited for similarity measures, since it organizes verbs, nouns, adjectives and adverbs with variation in semantic relations into synonym sets (synsets) by representing one concept. It uses *is-a* relation to organize noun and verbs into hierarchies. Semantic relations used by WordNet are autonomy, synonymy, member, hyponymy, domain, relation, cause and similar and so on. wup (Wu and Palmer 1994), lch (Leacock and Chodorow 1998) and $path$ calculates similarity with path length. lin (Lin 1998), res (Resnik 1995) and jcn (Jiang and Conrath 1997) measures similarity with information content, which is a corpus-based measure of the specificity of concept. WordNet also provides *is-made-of*, *has-part*, *is-an-attribute-of*, $etc.$, non-hierarchical relations. With this additional relations, measures of relatedness are also supported by WordNet which are $lesk$ (Banerjee and Pedersen 2003), hso (Hirst and St-onge 1998) and $vector$ (Patwardhan 2003).

6.3.4 Rocchio's Model

Rocchio's model [37] and Snippet Click model [36] are compared with RSR algorithm. Rocchio's Model [37] uses relevant and irrelevant URLs identified by users in search log to extend the input query. The extended query is used to carry out retrieval again. These URLs are converted into documents with title and snippet. Let the input query be q, the set of related documents accepted by users be D_r and the set of non-related documents be D_{ir}. The expanded query q_e is computed by using Eq. 6.5. Here, a, b and c are parameters and their traditional values are 1, 0.8 and 0.1, respectively. Related documents are given more importance than non-related documents. The importance of terms which are present in both related and non-related documents and only in non-related documents is reduced by subtraction.

$$q_e = aq + \frac{b}{\mid D_r \mid} \sum_{d_r \in D_r} d_r - \frac{c}{\mid D_{ir} \mid} \sum_{d_{ir} \in D_{ir}} d_{ir} \tag{6.5}$$

6.3.5 Snippet Click Model

Global-scale snippet click model [36] uses clicked URLs CLK_{url} from the user search log for a given input query q. Snippets are extracted for CLK_{url} and converted into documents D. Each keyword Term Frequency (TF) is calculated in documents D. Top N keywords with largest TFs is used as recommendation candidates. These N keywords are combined with the input query q and displayed as recommendations.

Related search recommendation framework is presented as shown in Fig. 6.1. Feedback sessions are generated for a given query from the user search logs and pseudo documents are mapped to it.

Feedback Sessions: Generally, a session can be defined as a list of consecutive queries to correlate a particular user search knowledge and clicked URLs for Web search [39]. Lu et al. [10] have focused on deriving a feedback session with a single query. In this chapter, query suggestions are generated for a query and hence a single session with a single query is suitable and is different from the traditional session.

The feedback session is defined with both clicked and un-clicked documents and it ends with last clicked documents in a session. This feedback session gives information that all the URLs have been examined and assessed by users before the last click. Figure 6.2 shows an example of feedback session for query *bank exam*. The left part is the 19 search results of the query *bank exam* and the right part is a user's click series, with 1 as clicked URLs by user and 0 as un-clicked. Here, a single session includes 19 URLs, while the feedback session includes only 15 URLs. The feedback session consists of four clicked and six un-clicked URLs. Inside this session, the clicked URLs display that is relevant to the users and the un-clicked URLs display that is irrelevant to the users. The un-clicked URLs followed by the last clicked URL are ignored in the feedback session since it is not assured that users have scanned or not.

Generate Enriched Documents from Feedback Sessions: It is not suitable to use feedback sessions directly to obtain meaningful information for suggestions as it may differ for different search history and queries. Usually, users have ambiguous keywords in their minds to represent their information need. Hence, it is not a good idea to generate relation between the user query keywords for recommendations. Enriched documents [10] are generated from feedback sessions and this enriched document is used to locate keywords that appear in snippets clicked and un-clicked documents in feedback session. The method of generating enriched document is given in Function 6.1.

Function 6.1: Enriched Document

Function: EnrichedDocument(FeedBack Session FS)

for *each URL u in Feedback Session* FS **do**

  Extract Title T and Snippet S

  Generate T_p from T after removal of stopwords, transforming all letters to lowercase and applying Stemming

  Generate S_p from S after removal of stopwords, transforming all letters to lowercase and applying Stemming

  Generate T_v and S_v vector by calculating Term Frequency-Inverse Document Frequency (TF-IDF) for T_p and S_p as shown in Eqs. 6.6 and 6.7 respectively

  Generate Enriched Document ED by the weighted sum of T_v and S_v as shown in Eq. 6.8

end

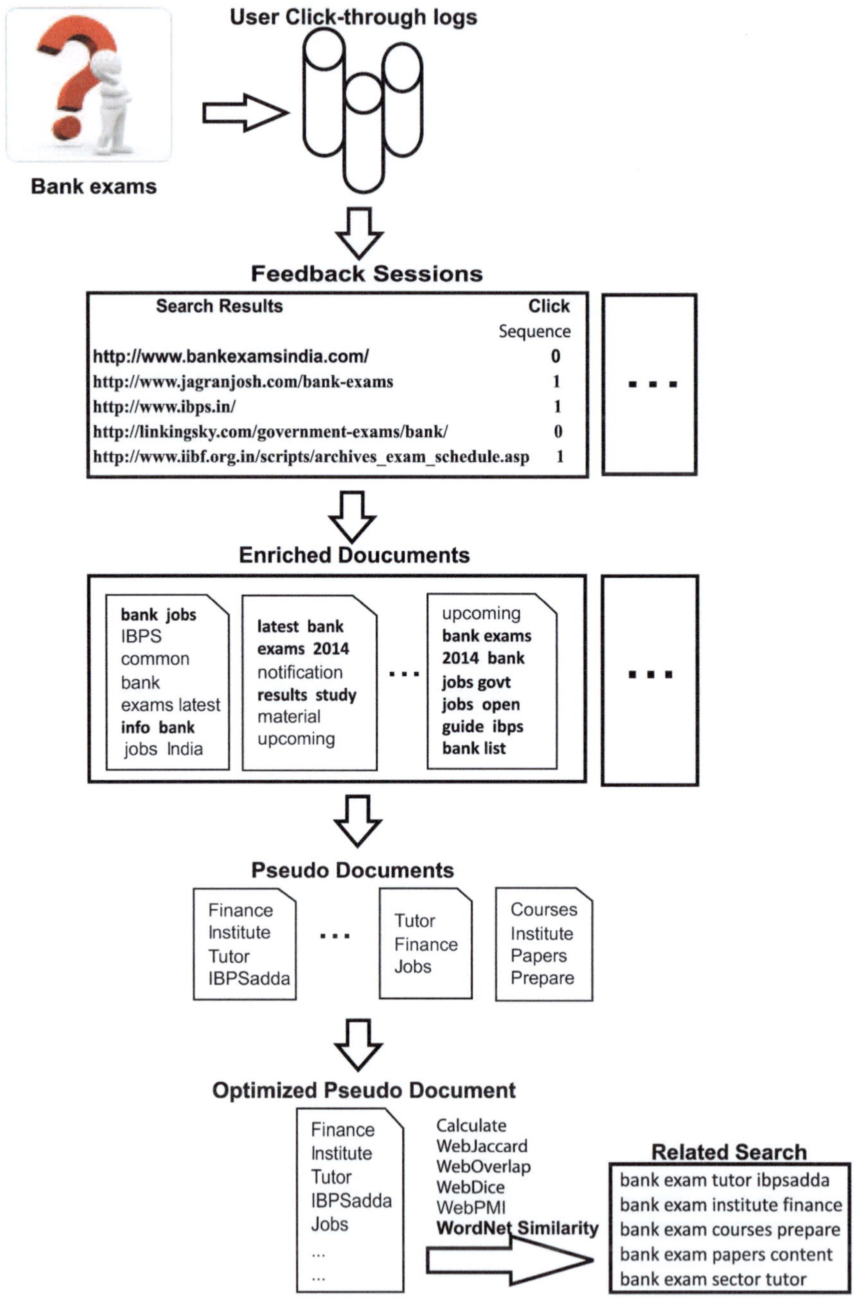

Fig. 6.1 Related search recommendation framework

Search Results	Click Sequence
Http://www.bankexamsindia.com/	0
http://www.jagranjosh.com/bank-exams	1
http://www.ibps.in/	1
http://linkingsky.com/government-exams/bank/	0
http://www.bankexamstoday.com/	0
http://www.freejobalert.com/upcoming-exam-dates-of-various-jobs/1835/	1
http://www.freejobalert.com/upcoming-notifications/21614/	0
http://www.time4education.com/bankexams/	0
http://www.bankjobsindia.net/upcoming-bank-exams-2014-in-india-and-latest-bank-jobs/	0
http://www.successcds.net/Bank-PO/	0
https://bankerschoice.talentsprint.com/	0
https://bankerschoice.talentsprint.com/indian-bank-exams-ibps-sbi/quant-formulae	0
https://www.sbi.co.in/user.htm?action=viewsection&lang=0&id=015110	0
http://www.bankingexamseasy.com/	0
http://www.iibf.org.in/scripts/archives_exam_schedule.asp	1
https://www.facebook.com/IbpsExamGuru	0
http://www.tcyonline.com/exam-preparation-bank-po-preparation-tests/100241/bank-po-clerical	0
http://www.eenadupratibha.net/Pratibha/ibps.aspx	0
http://www.sbirecruitment2014.org/	0

Fig. 6.2 An example of feedback session for query *bank exam* in rectangular box

T_v and S_v vectors are given in Eqs. 6.6 and 6.7.

$$T_v = [t_{w1}, t_{w2}, ...t_{wm}] \tag{6.6}$$

$$S_v = [t_{w1}, t_{w2}, ...t_{wn}], \tag{6.7}$$

where t_{wm} = Term Frequency–Inverse Document Frequency (TF-IDF) value of the mth term in URL's title and t_{wn} = TF-IDF value of the nth term in the URL's snippet. The enriched document is defined as given in Eq. 6.8.

$$ED = w_t T_v + w_s T_s = [ed_{w1}, ed_{w2} ....ed_{wk}] \tag{6.8}$$

where w_t is the weight of the title, w_s is the weight of the snippet and ed_{wi} indicates the importance of ith term in the URL. As the title directly represents the URL information, it is necessary to give more importance to title terms than the snippet terms, and therefore w_t is set to 2 and w_s is set to 1. Five enriched documents are generated for five URLs of feedback session (see Fig. 6.1).

Generate Pseudo Documents from Enriched Documents: For a feedback session, each URL is converted into enriched document. This document contains frequent terms that appears in clicked and un-clicked documents. For each feedback session, a Pseudo Document is generated from its enriched documents. The method of generating Pseudo Document (PD) is shown in Function 6.2.

Function 6.2: Pseudo Document

Function: PseudoDocument(FeedBack Session FS, Enriched Document ED)

for *each Feedback Session* FS **do**

Group Enriched Documents of FS, as $ED_{clk} = [ed_{w1clk}, ed_{w2clk}, ...ed_{wmclk}]$ and ED_{unclk}=$[ed_{w1unclk}, ed_{w2unclk}, ...ed_{wnunclk}]$ of the clicked and un-clicked URLs respectively.

for *each term in* $(ED_{clk}\ U\ ED_{unclk})$ **do**

Generate Pseudo Document PD by optimizing the value of term such that term t belongs to ED_{clk} gets more importance than t that belongs to ED_{unclk} as given in Eq. 6.9.

end

end

The generated $PD = [ed_{w1}, ed_{w2}...ed_{wp}]$

$$ed_w = \underset{ed_w}{\text{argmin}}\{\sum_M [ed_w - ed_{wclk}]^2 - \lambda \sum_N [ed_w - ed_{wunclk}]^2\} \tag{6.9}$$

Here, ed_w is the optimized term in Pseudo Documents, ed_{wclk} is the term from clicked enriched documents, ed_{wunclk} is the term from un-clicked enriched documents and λ is a parameter balancing the importance of clicked and un-clicked URLs. λ is set to 0.5 because if λ is set to a small value, then un-clicked URLs importance is reduced and if λ value is too large then un-clicked URLs dominates the value of ed_w. A pseudo document generated from five enriched documents is shown in Fig. 6.1.

Generate Optimized Pseudo Document from Pseudo Documents: The pseudo document reflects both the relevant and irrelevant documents to the users. Optimized Pseudo document is generated by combining all the pseudo documents for an input query. The method for generating optimized pseudo document is shown in Function 6.3. N is set to 10 as we observe that the top 10 terms are representing the users' information need.

Semantic similarity is calculated between optimized pseudo document terms by WebJaccard, WebDice, WebPMI, WebOverlap methods and WordNet-based similarity measures as discussed. Recommendation results are generated and ranked by combining query and terms for all the methods. These results are evaluated in performance evaluation.

6.3.6 RSR Algorithm

In this section, we present Related Search Recommendation (RSR) Algorithm as shown in Algorithm 6.1

Function 6.3: Optimized Pseudo Document

```
Function: OptPseudoDocument(Pseudo Document PD)
for each PD do
    select top N terms
    compute occurrence of each term in all the PDs
    Arrange the terms in descending order of occurrence and select top N terms for
    optimized PD
end
```

Algorithm 6.1: RSR: Related Search Recommendation

```
Input  : input query q, user click through log l
Output: related queries rq = < 1..k >
begin
    for input query q do
        Select Feedback Sessions FS = (fs_1, fs_2...fs_n) from user click through log l
        for each feedback session fs in FS do
            Generate enriched documents ED = (ed_1, ed_2...ed_m) by
            EnrichedDocument(FeedBack Session fs)
            Generate Pseudo document pd with PseudoDocument(Feedback session fs,
            enriched document ED)
            Add pd to PD < 1...l >
        Generate optimized pseudo document OPD = (opd_w1, opd_w2,....opd_wn) with
        OptPseudoDocument(Pseudocumets PD)
    for each opd_wi in OPD of size n do
        Calculate semantic similarity of opd_wi (1<i<n) AND opd_wj (i<j<n) with
        WebOverlap as discussed in section
        rq_Overlapi = q + opd_wi + opd_wj
    rq = rq_Overlapi
```

6.4 Experiments

6.4.1 Data Collection

To evaluate our proposed method, 95 students participated and each student is assigned 5 queries to collect the feedback session (Permission is taken from the Chairperson, Department of Computer Science and Engineering, UVCE, Bangalore). A Google middleware is implemented to monitor the user clicks. The top 50 search results from Google are retrieved for the submitted query. The title and web-snippets of resulting search are presented to the user as the snippets provide more

Table 6.2 Statistics of clicked information of users

Total users	95
Total queries allocated to each user	5
Total test queries	100
Total unique queries	100
Total URLs retrieved for a query	50
Total URLs retrieved	5000
Average feedback sessions for a query	5
Average clicked URLs for a query	9.732
Average un-clicked URLs for a query	40.268
Total words extracted from title for a query	23048
Average words extracted from title for a query	230
Total words extracted from snippet for a query	38098
Average words extracted from snippet for a query	380
Total words extracted	61146
Average words extracted for a query	611

information about the documents and help them to guide to the click URLs. Feedback sessions are generated through the clicked information of a user for a given input query. Table 6.2 shows the statistics of the clicked information of users for this experiment.

6.4.2 *Experimental Setup*

The setup of Related Search Recommendation (RSR) framework is as follows: Feedback sessions are generated for a given input query from the user click through *log* as discussed. Each URL in the feedback session is enriched with title and snippet terms after removing stopwords and applying stemming. Terms are weighed using Term Frequency–Inverse Document Frequency (TF-IDF) as explained in Function 6.1. Enriched documents of a feedback session are classified into clicked and un-clicked documents. Pseudo documents are generated by Eq. 6.9. Similarly, Pseudo documents are generated for all the feedback sessions for an input query. Optimized Pseudo document is generated by combining all the pseudo documents as shown in Function 6.3. Optimized Pseudo document has top-10 terms which reflect the user's information need. Semantic similarity between these terms t_s are calculated by WebJaccard, WebDice, WebPMI, WebOverlap methods and WordNet-based similarity measures. Recommendations are generated and ranked by combining query and terms t_s for all the methods.

The setup of Rocchio's model is as follows: User-identified relevant and irrelevant URLs are partitioned from the user click through *log* for a given input query. These URLs are converted into documents with title and snippet. Stopwords removal and stemming are applied for these documents to reduce noise. Expanded queries are generated by Eq. 6.5.

The setup of Snippet Click Model (SCM) is as follows: All the clicked URLs from user click through *log* are obtained for a given input query. Snippets are extracted from these URLs. Top-10 keywords are extracted by calculating the term frequency of the terms present in snippets. Query recommendations are generated by combining the input query with extracted keywords.

To examine the effectiveness of considering only clicked URLs in our proposed method (click-RSR), enriched documents are generated with only clicked URLs. Pseudo documents are generated by setting λ value to zero in Eq. 6.9 to remove the effect of un-clicked URLs. Optimized Pseudo document is generated by combining all the pseudo documents as shown in Function 6.3. Optimized Pseudo document has top-10 terms. Semantic similarity between these terms t_s are calculated by WebJaccard, WebDice, WebPMI and WebOverlap methods. Recommendations are generated and ranked by combining query and terms t_s for all the methods.

6.4.3 Query Recommendation Results

Top-5 recommendation results of Rocchio's model, Snippet Click model, Click-RSR and our RSR algorithm is shown in Table 6.3. Only terms are displayed in recommendation results due to space restriction. The actual recommendations for all models are query + terms. For query *bank exam*, recommendations for Rocchio's model are *bank exam finance*, *bank exam institute*, *bank exam tutor*, *bank exam ibpsadda* and *bank exam gr8ambitionz*. Recommendations for Snippet Click Model are *bank exam bank*, *bank exam competitive*, *bank exam exam*, *bank exam notification*, *bank exam awareness*. Recommendations for Click-RSR are *bank exam question bank*, *bank exam question tutor*, *bank exam papers bank*, *bank exam shortcuts bank*, *bank exam bank facebook*. Recommendations for the RSR algorithm are *bank exam tutor ibpsadda*, *bank exam institute finance*, *bank exam courses prepare*, *bank exam papers content*, *bank exam sector tutor*.

6.4.4 Performance Analysis

From the result shown in Table 6.3, it is observed that RSR algorithm recommends related queries to the given input query. Hundred test queries from various topics like Science, Shopping, Health care have been included.

Lu et al. [10] have discovered different users search goals for a query by using feedback session. This search goals can be utilized in query recommendations. Feedback

Table 6.3 Related search recommendation results comparison

Sr. no	Query	Rocchio's model [37]	Snippet click model [36]	Click-RSR	Proposed RSR algorithm
1	Bank exam	Finance	Bank	Question bank	Tutor ibpsadda
		Institute	Competitive	Question tutor	Institute finance
		Tutor	Exam	Papers bank	Courses prepare
		Ibpsadda	Notifications	Shortcuts bank	Papers content
		Gr8ambitionz	Awareness	Bank facebook	Sector tutor
2	Apartment	Budapest	Budapest	Budapest adina	Realestate properties
		Zillow	Apartment	Zillow rental	Realestate commonfloor
		Decor	Zillow	Zillow genuine	Realestate luxury
		Adina	123844	Rental genuine	Properties commonfloor
		Genuine	Luxury	Realestate properties	Properties luxury
3	Weather	Wiz	Forecast	History weather	BBC forecasts
		Kids	Weather	Wiz kids	BBC animated
		Welcome	Web	Wiz weather	Oceanic atmospheric
		Internet	Local	Web welcome	Forecasts australia
		Temperatures	Dallas	Web weather	Forecasts temperatures
4	Camera	Nokia	Sony	Nokia grip	Analog lense
		Android	Lines	Nokia 1020	Analog flash
		Pocket	Cameras	Nokia lumia	CCTV lense
		Grip	Nokia	Grip 1020	CCTV flash
		1020	Github	Grip lumia	Canon lense
5	Online recharge	Tariffs	MTNL	Payments cellone	Landline cellone
		Cellone	Prepaid	Recharge tariffs	State personal
		Personal	BSNL	Portal cellone	Landline postpaid
		State	Reliance	Prepaid tariffs	Landline huch
		Landline	Services	Banking personal	Landline packs
6	Free music	Jamendo	Music	Songza worthy	Downloads jango
		Songza	Appears	Composition notation	Downloads limewire
		Composition	Automated	Composition musescore	Beats freeplay
		Archive	Purple	Notation musescore	Beats uncopyrighted
		Jango	Listen	Streaming archive	Beats song
7	Scholar ships	Reimbursement	Pradesh	Reimbursement fee	Ph.D 2015
		Federal	IEEE	Federal applying	Ph.D postdoctoral
		Scholarshipporta	Scholarships	Federal finding	Ph.D masters
		Scholarshipexperts	Fellowships	Scholarshipportal solution	2015 postdoctoral
		Solution	Scholarship	Applying finding	2015 masters

(continued)

Table 6.3 (continued)

Sr. no	Query	Rocchio's model [37]	Snippet click model [36]	Click-RSR	Proposed RSR algorithm
8	Solar system	Youtube	Tour	Tour solar	Asteroids image
		Wikipedia	Ice	Youtube solar	Kidsastronomy image
		Meteorites	BBC	Youtube witness	Meteorites image
		Characteristics	Phet	Youtube peaceful	Views image
		Astronomy	Velocity	Youtube tucs	Visualizer image
9	Maths	Mathworld	Alpha	Mathworld Webs	Skills watch
		Ask	Wolfram	Mathworld wolfram	American homepage
		Webs	Puzzles	Webs wolfram	Youtube trick
		Level	Guardian	Ask forum	Youtube fast
		Extensive	Drexel	Warwick mathworld	Trick fast
10	Wedding	Facebook	ANN	Fairy tale	Blog cards
		Fairy	Pretty	Fairy disneys	Fairy tale
		Tale	Wedding	Tale disneys	Registries mywedding
		Disneys	Registry	Gifts fairytale	Blog etiquette
		Registries	Nordstrom	Nordstrom wedfolio	Blog popular

sessions are utilized in this work and the performance of RSR algorithm is compared with different recommendation methods like classical Rochhio's model [37], Snippet Click Model [36] and modified approach of RSR algorithm considering only clicked URLs. We have adopted Click Through Rate (CTR) method used in [36] to evaluate related search recommendations. CTR is the percentage of ever clicked recommendations in all recommendations for a given query. The set of students who participated in collecting click through log also participated in computing CTR as they can judge the recommendation results effectively. CTR is used to evaluate whether the recommendation is clicked by the user and a higher CTR value proves the effectiveness of the algorithm.

CTR is calculated for top-5 recommendations results generated with WebJaccard, WebDice, WebPMI and WebOverlap methods for RSR algorithm. The average value of CTR and ranked recommendations results are depicted in Fig. 6.3 for all the methods. The average CTR value of Top-5 recommendations are displayed in Table 6.4. CTR is also calculated for WordNet different semantic similarity measures. The average CTR value of Top-5 recommendations are displayed in Table 6.5. It is observed from WordNet similarity measures that few terms are not available in WordNet database, hence are not able to find out similarity between two terms. It is observed from Tables 6.4 and 6.5 that recommendations ranked with WebOverlap method have higher CTR value. Hence, WebOverlap method is adopted to rank RSR recommendations.

Similarly, CTR is calculated for top-5 recommendations results generated with WebJaccard, WebDice, WebPMI and WebOverlap methods for click-RSR algorithm.

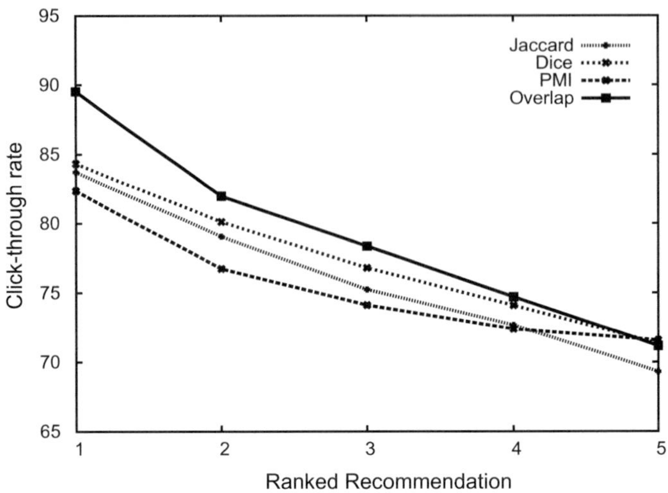

Fig. 6.3 CTR versus ranked recommendation results

Table 6.4 Average CTR value for Top-5 recommendations for RSR and Click-RSR algorithm

	WebPMI	WebJaccard	WebDice	WebOverlap
RSR	75.43	76.00	77.33	79.15
Click-RSR	73.72	72.72	71.20	74.02

Table 6.5 Average CTR value for Top-5 recommendations for WordNet similarity measures

lch	wup	path	res	lin	jcn	hso	lesk	vector
73.36	74.87	67.89	73.76	62.18	45.4	70.70	76.36	67.10

The average CTR value of Top-5 recommendations are displayed in Table 6.4. It is observed that recommendations ranked with WebOverlap method have higher CTR value. Hence, WebOverlap method is adopted to rank click-RSR recommendations.

To compare the RSR algorithm with other models, the average CTR value and ranked recommendations are displayed in Fig. 6.4. The average CTR value of Top-5 recommendations for all the models are depicted in Table 6.6. It is observed that the RSR algorithm has highest CTR value in comparison with other models.

It is observed that the CTR value of the RSR algorithm increases by 25% in comparison with SCM. The major difference between our algorithm and SCM is the consideration of un-clicked URLs along with clicked URLs, while SCM considers only clicked URLs. Even the weighing of terms in SCM is limited to term frequency which is further optimized in RSR algorithm.

The CTR value of the RSR algorithm increases by 24% in comparison with Rocchio's model. The difference between two approaches are as follows: (1) In our

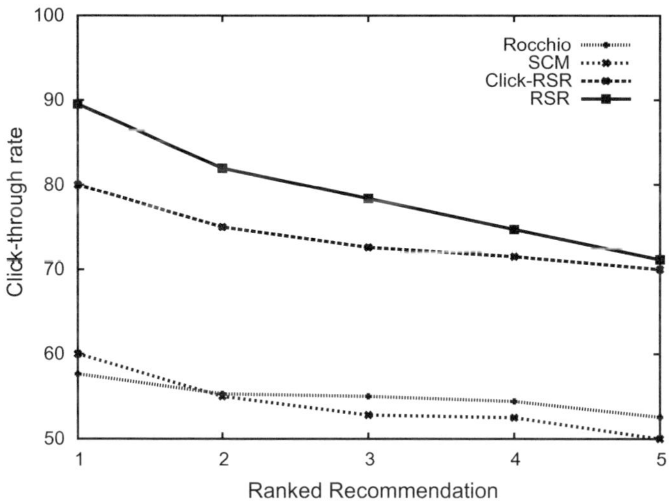

Fig. 6.4 CTR comparison with other models

Table 6.6 Average CTR value for Top-5 recommendations for all models

SCM [36]	Rocchio [37]	Click-RSR	RSR
54.06	55.03	73.82	79.15

approach, feedback sessions are limited to the last clicked URL as the left-out URLs may not be of user's interest. (2) Click through data is considered as sessions in RSR algorithm while in Rocchio's model it is treated as group of clicked/un-clicked URLs.

The CTR value of RSR algorithm increases by 5% in comparison with click-RSR. The major difference between RSR algorithm and click-RSR is the consideration of only clicked URLs in the feedback session. It is observed from the recommendations result from RSR algorithm that the terms from un-clicked URLs are also present. It is observed that top-5 recommendations from RSR algorithm for 100 test queries consists of about 23.5% of overall terms from the un-clicked URLs in the feedback sessions, which shows the importance of the un-clicked URLs scanned by users. Thus, the RSR algorithm outperforms the click-RSR.

6.5 Summary

In this chapter, we have presented Related Search Recommendation (RSR) algorithm to suggest related queries to given input query by using feedback session from user click through log. Each feedback session is converted into enriched documents. Pseudo Documents are generated by combining all the enriched documents of a feedback session. Optimized Pseudo Document is generated by combining all the Pseudo Documents for a given input query, which reflects the user's information need. Semantic similarity is calculated by WebJaccard, WebDice, WebPMI and WebOverlap methods for terms present in the optimized Pseudo Document. Recommendations are generated and ranked by combining query and terms for all the methods. Simulations are performed on click through log generated by displaying title and snippet to the students of our college and compared with Rocchio's model, Snippet Click Model and Click-RSR. Click Through Rate (CTR) is used as a performance evaluation metric. Simulation results show that RSR algorithm outperforms Rocchio's model, Snippet Click Model and Click-RSR by providing higher CTR value. Further, this work can be extended to classify the search results into different topics.

References

1. M.P. Kato, S. Tetsuya, T. Katsumi, When do people use query suggestion? A query suggestion log analysis. Journal of Information Retrieval, 16(6), 725–746 December (2013)
2. S. Craig, M. Hannes, H. Monika, M. Michael, Analysis of a very large web search engine query log, in *SIGIR Forum* (1999), pp. 6–12
3. C. Huanhuan, J. Daxin, P. Jian, H. Qi, L. Zhen, C. Enhong, L. Hang, Context-aware query suggestion by mining click-through and session data. KDD '08, in *The Proceedings of 14th International ACM SIGKDD Conference on Knowledge Discovery and Data Mining* (2008), pp 875–883
4. M. Qiaozhu, Z. Dengyong, C. Kenneth, Query suggestion using hitting time, in *The Proceedings of 17th ACM Conference on Information and Knowledge Management, CIKM '08* (2008), pp. 469–477
5. M. Hao, M.R. Lyu, I. King, Diversifying query suggestion results, in *The Proceedings of 24th AAAI International Conference on Artificial Intelligence, AAAI '10* (2010), pp. 1399–1404
6. J. Guo, X. Cheng, G. Xu, H. Shen, A structured approach to query recommendation with social annotation data, in *The Proceedings of 19th ACM International Conference on Information and Knowledge Management, CKIM '10* (2010), pp. 619–628
7. Y. Song, D. Zhou, L.-W. He, Post ranking query suggestion by diversifying search results, in *The Proceedings of 34th International ACM SIGIR Conference on Research and Development in Information Retrieval, SIGIR '11* (2011), pp. 815–824
8. Z. Kunpeng, W. Xiaolong, L. Yuanchao, A new query expansion method based on query logs mining. Int. J. Asian Lang. Process. **19**(1), 1–12 (2009)
9. Y. Chen, Y.-Q. Zhang, A personalized query suggestion agent based on query-concept bipartite graphs and concept relation trees. Int. J. Adv. Intell. Parad. **1**(4), 398–417 (2009)
10. Z. Lu, H. Zha, X. Yang, W. Lin, Z. Zheng, A new algorithm for inferring user search goals with feedback sessions. IEEE Trans. Knowl. Data Eng. **25**(3), 502–513 (2013)
11. J. Miao, J.X. Huang, Z. Ye, Proximity based Rocchios model for pseudo relevance feedback, in *The Proceedings of 35th ACM International Conference on Research and Development in Information Retrieval, SIGIR' 12* (2012), pp. 535–544
12. M.S. Hamani, R. Maamri, Word semantic similarity based on document's title, in *The Proceedings of 24th IEEE International Workshop on Database and Expert Systems Applications* (2013), pp. 43–47

13. D. Bollegala, Y. Matsuo, M. Ishizuka, Measuring semantic similarity between words using web search engines, in *The Proceedings of 16th International Conference on World Wide Web, WWW '07* (2007), pp. 757–766
14. P. Li, H. Wang, K.Q. Zhu, Z. Wang, X. Wu, Computing term similarity by large probabilistic *isA* knowledge, in *The Proceedings of 22nd International Conference on Information and Knowledge Management, CIKM '13* (2013), pp. 1401–1413
15. B. Danushka, M. Yutaka, I. Mitsuru, A relational model of semantic similarity between words using automatically extracted lexical pattern clusters from the web, in *The Proceedings of International Conference on Empirical Methods in Natural Language Processing, EMNLP '09* (2009), pp. 803–812
16. B. Danushka, M. Yutaka, I. Mitsuru, A web search engine-based approach to measure semantic similarity between words. IEEE Trans. Knowl. Data Eng. **23**(7), 977–990 (2011)
17. P. Resnik, Using information content to evaluate semantic similarity in a taxonomy, in *The Proceedings of the 14th International Joint Conference on Artificial Intelligence* (1995), pp. 448–453
18. D. Lin, An information theoretic definition of similarity, in *The Proceedings of the Fifteenth International Conference on Machine Learning* (1998), pp. 296–304
19. C. Leacock, M. Chodorow, Combining local context and wordnet similarity for word sense disambiguation, in *WordNet: An Electronics Lexical Database* (1998), pp. 265–283
20. J.J. Jiang, D.W. Conrath, Semantic similarity based on corpus statistics and lexical taxonomy, in *The Proceedings of International Conference on Research in Computational Linguistics* (1997), pp. 19–33
21. E. Agirre, E. Alfonseca, K. Hall, J. Kravalova, M. Paşca, A. Soroa, A study on similarity and relatedness using distributional and wordnet-based approaches, in *The Proceedings of Human Language Technologies: The 2009 Annual Conference of the North American Chapter of the Association for Computational Linguistics* (2009), pp. 19–27
22. G. Hirst, D. St-Onge, Lexical chains as representations of context for the detection and correction of malapropisms, in *WordNet: An Electronics Lexical Database* (1998), pp. 305–332
23. Y. Song, L.-W. He, Optimal rare query suggestion with implicit user feedback, in *The Proceedings of 19th International Conference on World Wide Web, WWW '10*, pp. 901–910 (2010)
24. N. Craswell, M. Szummer, Random walks on the click graph, in *The Proceedings of 30th Annual International ACM SIGIR Conference on Research and Development in Information Retrieval, SIGIR '07* (2007), pp. 239–246
25. F. Radlinski, T. Joachims, Query chains: learning to rank from implicit feedback, in *The Proceedings of 11th ACM SIGKDD International Conference on Knowledge Discovery in Data Mining, KDD '05* (2005), pp. 239–248
26. G.-R. Xue, H.-J. Zeng, Z. Chen, Y. Yu, W.-Y. Ma, W. Xi, W. Fan, Optimizing web search using web click through data, in *The Proceedings of 13th ACM International Conference on Information and Knowledge Management, CIKM '04* (2004), pp. 118–126
27. E. Kharitonov, C. Macdonald, P. Serdyukov, L. Ounic, Intent.estions, in *The Proceedings of 22nd ACM Conference on Information and Knowledge Management, CIKM '13* (2013), pp. 2303–2308
28. U. Ozertem, O. Chapelle, P. Donmez, E. Velipasaoglu, Learning to suggest: a machine learning framework for ranking query suggestions, in *The Proceeding of the 35th International ACM SIGIR Conference on Research and Development in Information Retrieval, SIGIR '12* (2012), pp. 25–34
29. D. Broccolo, L. Marcon, F.M. Nardini, R. Perego, S. Fabrizio, Generating suggestions for queries in the long tail with an inverted index. Int. J. Inf. Process. Manag. **48**(2) (2012)
30. X. Zhang, S. Zilles, R.C. Holte, Improved query suggestion by query search, in *The Proceedings of the 35th Annual German Conference on Artificial Intelligence* (2012), pp. 205–216
31. E. Gomez-Nieto, R.F. San, P. Pagliosa, W. Casaca, E.S. Helou, M.C.F. de Oliveira, L.G. Nonato, Similarity preserving snippet-based visualization of web search results. IEEE Trans. Vis. Comput. Graph. **20**(3), 457–470 (2014)

32. X.-H. Phan, C.-T. Nguyen, D.-T. Le, L.-M. Nguyen, S. Horiguchi, Q.-T. Ha, A hidden topic-based framework toward building applications with short web documents. IEEE Trans. Knowl. Data Eng. **23**(7), 961–976 (2011)
33. Q. He, D. Jiang, Z. Liao, S.C.H. Hoi, K. Chang, E.-P. Lim, H. Li, Web query recommendation *via* Sequential Query Prediction, in *The Proceedings of IEEE 25th International Conference on Data Engineering* (2009), pp. 1443–1454
34. Q. Jiang, M. Sun, Fast query recommendation by search, in *AAAI Conference on Artificial Intelligence* (2011), pp. 1192–1197
35. L. Li, G. Xu, Z. Yang, P. Dolog, Y. Zhang, M. Kitsuregawa, An efficient approach to suggesting topically related web queries using hidden topic model. Int. J. World Wide Web **16**(3), 273–297 (2013)
36. Y. Liu, J. Miao, M. Zhang, S. Ma, L. Ru, How do users describe their information need: query recommendation based on snippet click model. Int. J. Expert. Syst. Appl. **38**(11), 13874–13856 (2011)
37. J.J. Rocchio, Relevance feedback in information retrieval, in *The SMART Retrieval System: Experiments in Automatic Document*, ed. by G. Salton (1971), pp. 313–323
38. C. Fellbaum, *WordNet: An Electronic Lexical Database. Language, Speech, and Communication*. MIT Press, Cambridge (1998)
39. R. Jones, K.L. Klinkner, Beyond the session timeout: automatic hierarchical segmentation of search topics in query logs, in *ACM Conference on Information and Knowledge Management* (2008), pp. 699–708

Chapter 7
Web Page Recommendations Based Web Navigation Prediction

K. R. Venugopal and Sejal Santosh Nimbhorkar

Abstract A huge amount of user request data is generated in Web log. Predicting users' future requests based on previously visited pages is important for Web page recommendation, reduction of latency and on-line advertising. These applications compromise with prediction accuracy and modelling complexity. In this chapter, a Web Navigation Prediction Framework for Web page Recommendation (WNPWR) which creates and generates a classifier based on sessions as training examples is proposed. As sessions are used as training examples, they are created by calculating the average time on visiting Web pages rather than the traditional method which uses 30 min as default time-out. The proposed method uses standard benchmark datasets to analyse and compare our framework with two-tier prediction framework. Simulation results show that our generated classifier framework WNPWR outperforms two-tier prediction framework in prediction accuracy and time.

7.1 Introduction

A huge amount of data is generated when millions of users access Websites. One of the influential data source is log file of Web server, which traces the users' web-browsing actions. This data consists of repeatedly accessed Web pages by users within a period of time. Users' navigation history within a period of time is known as session. This session information is very important and helpful to find the user behaviour. Through this behaviour, user's next request can be predicted and recommendations can be made to reduce the browsing time of the Web pages. Recommending related pages to users reduces network traffic as it avoids visiting unnecessary pages.

The possibility of visiting a Web page by a user based on the history of earlier accessed Web pages is known as Web prediction. Prediction of Web user's behaviour

K. R. Venugopal (✉)
Bangalore University, Jnana Bharathi, Bengaluru 560056, India
e-mail: venugopalkr@gmail.com

S. Santosh Nimbhorkar
BNM Institute of Technology, Banashankari, Bengaluru 560070, India
e-mail: sej_nim@yahoo.co.in

K. R. Venugopal et al., *Web Recommendations Systems*,
https://doi.org/10.1007/978-981-15-2513-1_7

is critical in Web mining to enhance the performance of the search engine. The organization of the Web is modelled as a graph, where each node represents a Web portal, and the edge represents the user's navigation. Distribution of all Web pages visited can be calculated and utilized in re-weighting and re-ranking results. The information provided by the navigation path is of prime importance than the query given by the user. Web cache performance of search engine can be improved by storing predicted pages in the cache.

The Web users' usually spend more time on browsing and authoring than on search. Hence, the search engine cannot effectively predict users' search intention. Prediction is performed only after the users submit their queries to search engines, as the prediction is conducted in a passive manner and this navigation history is used for Web prediction. Web prediction can be used in recommendation system, in which the top k users are involved in similar activity.

Behavioural targeting is a key issue of predicting future behaviour of Web users. Behavioural targeting is a technique to improve efficiency of advertising by online website publishers and advertisers by extracting knowledge of web-browsing behaviour of users. Behaviour targeting selects advertisement to display with the help of web-browsing behaviour of users. The user analysis approach is the centre of interest in on-line advertising and properly targeted advertisements generate more consumer interest.

Predicting Web user's shopping behaviour has an important role in product recommendation. Product recommendations are the dynamic shopping recommendations across mobile, email and Web channels. It depends on each customer's past and current purchase behaviours. It also helps in website optimization, improves conversions and increases revenue by making related product recommendations to the customers.

The World Wide Web (WWW) has created tremendous opportunities to spread and accumulate huge online information. This motivates researchers to understand the navigation behaviour of Web site visitors from Web usage data to improve the quality of service of that site, to reduce access latency and Web page recommendations using efficient Web prediction technique. The Web log records the navigational behaviour of the user. Preprocessing of the raw data is required before giving the data as input to prediction model. Preprocessing challenges include session identification, handling huge amount of data and obtaining domain intelligence. Low accuracy and expensive training are fundamental issues in prediction.

Many researchers have developed several prediction models by fusing Support Vector Machine (SVM), Markov model, Association Rule Mining (ARM) and Artificial Neural Network (ANN). First, SVM and Markov improve prediction time, but this model uses the traditional method of session identification with 30 min time-out period for one session which decreases the prediction accuracy. Second, SVM and ARM are not scalable with large datasets. Third, ANN and SVM cannot handle the multi-class issue effectively on account of a large number of classes that are used in Web prediction.

Web navigation prediction framework for Web page recommendation is designed by using user request on the Web. This framework creates a classifier based on sessions as training data and classifiers generated by the Nth-order Markov models.

Each session is mapped to the generated classifiers. If any session is mapped to more than one classifier, then each session is mapped to only one classifier according to PageRank algorithm during the filtering process. Filtered data is then used to train the SVM classifiers. Once SVM classifiers are trained, prediction accuracy and time are calculated on the test dataset. Finally, SVM classifiers can be used for page recommendation. In this work, Sessions are created by computing the average time on visiting Web pages rather than the traditional method which uses 30 min as default time-out. The PageRank algorithm is used in the filtering process, which results in an improvement in prediction accuracy and time.

7.2 Related Works

In this section, we have reviewed several papers related to Web page prediction and various prediction application.

7.2.1 Web Page Prediction

7.2.1.1 Markov Models for Web Page Prediction

Chimphlee et al. [1] developed a prediction method to access websites with First-order and Second-order Markov model by considering user Web access behaviour log file. This algorithm is used to cluster similar transition behaviours to further improve the efficiency of prediction. The First-order Markov model persistently gives the best performance to predict Web access behaviour. When the recall is less than 50%, the Second-order Markov model gives the worst performance. When precision is less than 50%, the association rule gives the worst performance. Next, Borges et al. [2] developed a method to measure variable-length Markov model's ability for summarizing Web sessions of users for the given length. Spearman footrule metric is used to determine the accuracy to characterize information content of the users' sessions. A prediction algorithm which eliminates few states of all kth Markov model selectively is developed to predict user's Web page access behaviour [3]. Support, Confidence and Error pruning technique are used to eliminate states. This technique has achieved better prediction accuracy than First, Second, Third and All kth Markov model.

7.2.1.2 Online Prediction

Guerbas et al. [4] proposed an approach for online navigational pattern prediction. Navigational patterns are discovered by density-based algorithm. A model is developed by modelling user's Web access information and by constructing weighted suffix tree from content of Web page [5]. This method requires less memory space and consistent computational training for user's activity. A two-step prediction model is developed that decreases the size of Web pages' candidate set and increases prediction accuracy's speed by using a hierarchical property of website [6].

7.2.1.3 Statistical Theory for Web Page Prediction

Many methods are developed to understand web-browsing behaviour using statistical theory. Dembczynski et al. [7] described the user-level models to predict Web users' behaviour by Statistical Decision Theory and Learning Theory. A model is proposed to understand web-browsing behaviour through Weibull distribution on *dwell time* [8]. *Dwell time* is the length of time a user spends on document. The log data is organized into sessions, each of which is determined as a sequence of Web pages browsed for 30 min or user closes browser before 30 min. *Dwell time* is computed for each page by leaving the last page in the session and Weibull distribution is applied to it. The prediction model is used to predict Web page *dwell time* distribution. Negative binomial distribution and inverse Gaussian models are used for qualitative comparison of session length to model the behaviour of visitors to an academic website [9].

White et al. [10] presented a log-based study to model user interests while interacting with the Web. The current page usage with other information like recent correspondence behaviour, hyperlinks, pages' relation to the present page which shares similar search engine queries, long-term interests of the present user with other users who also visit the present page are evaluated. Thwe [11] proposed a popularity- and similarity-based PageRank algorithm to predict Web page access behaviour. Different navigational attributes like size of the page, transition, frequency of page, similarity of the page, duration of the page and access time of the page are used. This model for next page prediction is a promising approach than Markov models.

7.2.1.4 Hybrid Techniques for Web Page Prediction

Different hybrid techniques are developed for Web page prediction. Khali et al. [12] presented a novel approach by incorporating association rules, Markov models and clustering for Web page prediction. The integration provides better prediction accuracy than using each technique individually. Awad et al. [13] studied composite models by combining various classification methods especially Artificial Neural Networks (ANN), All *k*th Markov Model by Dempster's rule and Markov Model to predict user's future request. Markov model lacks in predicting user's future request when training data is not available. In ANN, the prediction accuracy decreases when the number of classes increases. Markov model and ANNs combination handles above-mentioned drawbacks of individual model. This hybrid model is more effective in prediction accuracy than All *k*th Markov model, association rule mining and Markov model.

Dutta et al. [14] proposed a Web page prediction model by clustering user's interest and Markov model. Similar pages are aggregated with K-medoids clustering method and K is computed with the HITS algorithm. The predicted Web pages are saved with cellular automata scheme and are memory efficient. Awad et al. [15] developed a two-level prediction framework to identify user's web-browsing behaviour. Sessions are trained with nth-order Markov model and mapped to one or more orders of Markov models. Support Vector Machine (SVM) is used as a prediction technique to create Example Classifier (EC). A testing example is given as an input to EC to predict an appropriate classifier. This model predicts better web-browsing behaviour than Markov model and association rule mining.

7.2.2 Prediction Applications

Web Prediction is used in many applications like mobile users' movement, place prediction, service recommendations, online user behaviour, search prediction and image click prediction. Tseng et al. [16] designed a method to discover mobile users' sequential patterns to understand their movement in correlation with a desired service. Sequential Mobile Access Pattern (SMAP) tree is constructed to aggregate the access pattern. SMAP-Mine algorithm based on depth-first search is applied to find sequential patterns. A graph is created from session to understand behaviour strategy in mobile Internet, in which each Web page is vertex and edge indicates the number of transitions from one Web page from another Web page [17]. Random walk restart algorithm is applied on graph for prediction.

Semantic Place Prediction is a process to predict the semantic meaning of place. Huang et al. [18] have proposed a novel prediction framework which takes into account the spatial property, users' behaviour and environment for semantic place prediction. Several models like SVM, J48, etc., are used to build multilevel classification models. Decision Tree is used to discover the association between the results of different models and the real answer of place. Service recommendation system predicts the availability to atomic Web services by service Load, User Location, Service Class, Service Location Model [19]. This model predicts atomic services with collaborative filtering algorithm by using prior work availability. It predicts service availability by the geographic location of the service, the users' geographic location, the computational requirements of the service and the current load of the service provider.

Huang et al. [20] studied the usage of parallel browsing with the help of Web log. The degree of parallel browsing is identified by discovering browser pageview for outgoing clicks and tab switches. It is observed that 57.4% sessions with tab include parallel browsing and users are separating their browsing movement into various tabs rather than examining more pages. Goel et al. [21] developed a method to measure online behaviour of a Web user. This method demonstrates behavioural changes of the Web user with respect to time spent online. The heaviest users allot twice as much of their time to social media relative to a typical Web user. Linear Support Vector Machines(SVMs) is used to infer demographic attributes from the browsing history.

Cheng et al. [22] presented a technique to predict search intent from the user browsing behaviours. Queries are extracted from the Web pages that users read after issuing query from user browsing behaviour. Page-query bipartite graph is constructed and query visibility, query popularity and pattern frequency features are extracted from the graph to describe the users' interests. Query dissimilarity measure is also obtained from the bipartite graph to minimize the diversification of queries. Ranked list of the queries is obtained by Support Vector Machine (SVM) and suggest to users. Zhang et al. [23] designed task-centric click model to predict the user search behaviour. The sequence of queries and their clicks in search session is considered as a task. This model describes the user behaviour associated with a task as a collective

Table 7.1 A comparison of related works

Author	Concept	Advantages	Disadvantages
Awad et al. [13]	Predicting User's future request by combining Markov model and Artificial Neural Networks (ANNs)	This hybrid model is more effective in prediction accuracy than All kth Markov model, association rule mining Markov model	Requires more computation for prediction and training
Chimphlee et al. [1]	Predicting Web site access with user browsing history by hybrid Markov model	The 1st order Markov model constantly provides excellent prediction performance and model building is very easy	A particular order of Markov model is not able to predict for a session which was not examined during training because such session will have zero probability
Goel et al. [21]	Measures online browsing behaviour of a Web user by linear support vector machines (SVMs) from browsing history	Accurately measures individual activity with large scale data	SVM do not handle the multi-class problem efficiently
Awad et al. [15]	Identify user's web-browsing behaviour using nth order Markov model and support vector machine	Increases prediction accuracy and reduces prediction time compared to Markov and association rule mining models	Statistical features are not used to create sessions
WNPWR	Web navigation prediction framework for Web page recommendation	Enhances prediction accuracy and reduces prediction time compared to Awad et al. [15] method	

whole and shows improvement in prediction over the User Browsing Model and Dynamic Bayesian Networks.

Tian et al. [24] proposed a method which automatically predicts the feature of image search results for a query. First, features to measure image search quality is derived by examining the visual distribution attributes of bad and good search results of the training queries. The latent relationship between obtained features and the inherent query difficulty is mined during learning process to build query difficulty prediction model. This model is used to measure query difficulty for a new query. A method is designed to predict an image click based on hypergraph learning-based sparse coding method [25]. The acquired click data is used to re-rank images. Based on a group of the Web images with associated clicks and a new image without any clicks known as codebook, sparse coding is utilized to choose a few basic images as possible from the codebook in order to linearly reconstruct a new input image while minimizing reconstruction errors. A voting strategy is utilized to predict the click as a binary event from the sparse codes of the corresponding images.

Table 7.1 shows the comparison of closely related works with our proposed method.

7.3 Web Navigation Prediction Framework and WNPWR Algorithm

7.3.1 Problem Definition

Given a user navigation history, we convert user navigation history into sessions by calculating the average time of visiting Web pages. The objective is to generate Web navigational prediction framework with high prediction accuracy and reduced prediction time. The prediction accuracy and time is calculated only for the first four pages visited by the users. If the user visits more than four pages, then a sliding window of size four is applied.

7.3.2 Session Identification Method with Average Time of Visiting Web Pages

Session identification method with the average time of visiting Web pages that is used to create sessions to Web navigation prediction frameworks. Often the visitors access the same website frequently that is recorded in the log file. The process of segregating the page access of each user into a singular session is called session identification. It is presumed that the user starts a new session if the consecutive page request exceeds a certain time limit. Commercial websites usually have a default time-out of 30 min [26]. However, this time-out period may not be enough for some websites in which user reads articles and collects opinions about products. The time required to cover a certain amount of information is dependent on the profile of the users, for example, an elderly person follows information slowly. When a client wishes to purchase a product then he may spend more time to analyse the product and may exceed the 30 min time-out.

Dinuca et al. [27] proposed a new session identification method by computing the average time of visiting Web page. For each visited page, the duration of visit is computed as a difference between two successive timestamps for the same user and is identified either by username or IP. The highest timestamp among those visited pages by an user is assumed to be 20,000 s. A page average visit time is computed by calculating the mean of all the visit time spent on page. Time less than 2 s and larger than 20,000 s are not considered for computing the average visiting time. The method for session identification to compute the average time of visiting Web pages is given in Function 7.1.

Function 7.1: Session Identification

Function: session

Data: Consider the set of users by $U = U_1, U_2, ..., U_n$. The pages visited by the users U_k is identified by $PU_k = PU_{k1}, PU_{k2}, ...$ and $TSPU_{ki}$ is the timestamp of PU_{ki} page. $IDPU_{ki}$ is the session identification number allocate to pages PU_{ki} with ID.

for *each* U_k *in U Repeat* **do**
    $IDPU_{k1} = \max(ID)+1$;
    I = 1;
    while $(I < |PU_k|)$ **do**
        I=I+1;
        $IDPU_{ki} = IDPU_{k,i-1}$;
        $TMA_{ki} = \max(2 * TM_{ki}, 300)$;
        if $TSPU_{ki}$ - $TSPU_{k,i-1} > TMA_{ki}$ **then**
            $IDPU_{ki} = IDPU_{k,i-1} + 1$;

TM_{ki} is the average spent time on the page PU_{ki} by the users. TMA_{ki} is the time utilized to create sessions instead of 30 min time-out. The value of 300 is required to calculate TMA_{ki}, as the average time of some pages is very low that can negatively affect to identify the sessions.

For example, if the user X has visited different Web pages in a Web log as shown in Table 7.2. The visited Web pages are arranged in ascending order based on timestamp and duration of visited Web page is calculated as a difference between the timestamp of two successive Web pages as shown in Table 7.3. As discussed earlier, the timestamp difference of Web page with highest timestamp is assigned to 20,000. If the user Y has visited Web page B in his session and the visit duration is 40, then the average visit time of page B is $(53 + 40)/2 = 46.5$. Sessions are generated as described in Function 7.1.

Table 7.2 Web pages visited by a user

Web pages	Timestamp
A	[01/Jul/1995:01:40:52 -0400]
B	[01/Jul/1995:01:41:43 -0400]
C	[01/Jul/1995:01:42:36 -0400]
D	[01/Jul/1995:01:49:23 -0400]

Table 7.3 Timestamp difference between successive Web pages

Web pages	Timestamp difference
A	51
B	53
C	407
D	20000

7.3.3 Prediction Models

The Markov model, PageRank algorithm and Support Vector Machine (SVM) are used to generate classifiers in prediction framework.

Markov Model: Markov model is a stochastic process in which the next state relies on the former states. In Web prediction, the next state correlates to predicting the next visiting page and the former states correlate to the previously visited pages. Markov models are defined by three parameters, namely $< P, S, T >$ in Web prediction, where P is the previously visited Web pages by the users, S is the all possible states to build the Markov model; and T is a $|S| \times |P|$ Transition Probability Matrix (TPM) in which each entry t_{ij} represents the probability that a user visits page j when he has already visited i pages [3].

The simplest Markov model also known as first-order Markov model predicts the next page by only observing the previously visited page by the user. In the first-order model, the states represent a single page; in the second-order models, the states represent two successive pages and so on. In general, the Kth-order Markov model computes the probability of user visits kth page after he has visited $k - 1$ pages.

After determining the states of the Markov model, the TPM is estimated. The general approach is to use sessions as training set and measure each t_{ij} entry with the visited pages' frequency. For example, consider the session SE_6 (A, C, F, G, H) shown in Table 7.4. For the first-order Markov model, each state corresponds to a single page, so the first page A correlates to the state s_1 and second page C correlates to the state s_3. Since page A pursues the state s_3, the value of t_{13} in the TPM is amended. Equivalently, the next state is s_6 and the entry t_{36} is updated in TPM. Table 7.5 shows first-order Markov model TPM entries. In the higher order Markov model, each state is formed with more than one page. In the second-order Markov model, for session SE_6 the first state is (A, C) and the page F pursues the

Table 7.4 Session data

Session	Visited pages
SE_1	(A, B, C, D, E)
SE_2	(A, B, C, D, F)
SE_3	(A, B, C, D, E)
SE_4	(A, B, C, D, F)
SE_5	(A, B, F, G, H)
SE_6	(A, C, F, G, H)
SE_7	(A, B, C, D, E)

Table 7.5 First-order Markov model

1st order	A	B	C	D	E	F	G	H
$S_1 = \text{A}$	–	6	1	–	–	–	–	–
$S_2 = \text{B}$	–	–	5	–	–	1	–	–
$S_3 = \text{C}$	–	–	–	5	–	1	–	–
$S_4 = \text{D}$	–	–	–	–	3	2	–	–
$S_5 = \text{E}$	–	–	–	–	–	–	–	–
$S_6 = \text{F}$	–	–	–	–	–	–	2	–
$S_7 = \text{G}$	–	–	–	–	–	–	–	2
$S_8 = \text{H}$	–	–	–	–	–	–	–	–

state (A, C), the TPM value equivalent to the state (A, C) and page F is amended. Markov model has two advantages: (i) model construction efficiency and (ii) better prediction time performance. In our framework, we have used first-, second-, third- and fourth-order of Markov models.

PageRank Algorithm: Brin [28], designed a PageRank algorithm that ranks pages returned by a search engine. It allocates a numerical value to Web pages to compute their corresponding position in the Web pages' set. The significance of a page is equivalent to the total significant scores of Web pages linked to it. *PageRank* of a given page is the number of times the user has accessed the given page divided by the total number of pages the user has visited. The method for *Page Rank* is given in Function 7.2.

For example, if user X has visited different Web pages in a session as shown in Table 7.4, then pagerank of the page A is $7/8$, page B is $6/8$, page E is $3/8$ and page F is $5/8$.

Support Vector Machine (SVM): SVMs are used for classification which identifies patterns based on statistical learning theory. A classification method commonly separates data in testing and training sets. Each example in the training set includes one *target value* (i.e. the class labels) and various *attributes* (i.e. the observed or features variables). Given a set of training instances, each is assigned to one of the two categories. An SVM training algorithm generates non-probabilistic binary linear

Function 7.2: Page Rank

Function: PageRank
Data: Consider the set of users $U = U_1, U_2,...,U_n$. The pages visited by the user is identified by $PU_k = PU_{k1}, PU_{k2},...,PU_{km}$. Page rank of page PU_{ki} is $PRPU_{ki}$.

for *each* U_k *in* U *repeat* **do**
  for *each page in* PU_k *repeat* **do**
    Let $visit_{pages}$ = number of times the U has visited given page (PU_{ki})

$$PRPU_{ki} = \frac{visit_{pages}}{\mid PU_k \mid}$$

classifier which designates new instance to one of the categories. An SVM model represents instances as points in space and mapped in such a way that the instances with different categories are divided by a clear wide gap. New instances are then outlined into that same space and used for prediction. It avoids the curse of dimensionality problem and works well with high-dimensional data. The method to find SVM model is as shown in Function 7.3.

Function 7.3: Support Vector Machine

Function: SVM
Data: Consider Data instances with Class Label C_{label} and Attributes $C_{attributes}$

begin
  Segregate data instances into Training and Testing Dataset
  Convert each data instance as a vector of real numbers
  Apply scaling on Training and Testing Datasets
  Select SVM Kernel
  SVM produces a model SVM_{model} which predicts the target values
  Give test data attribute as input to SVM_{model} to predict the target value
  Use target value for recommendation

In our framework, the first-, second-, third- and fourth-order of Markov prediction models are used as data instances with predicted page as C_{label} and sessions as $C_{attributes}$. Here, LIBSVM package is used for SVM implementation and is discussed in experiments.

7.3.4 Two-Tier Prediction Framework

A two-tier prediction framework is discussed that is compared with our framework, as this framework has also used Markov model and SVM. Awad et al. [15] developed two-tier prediction framework to identify user's web-browsing behaviour. Sessions are trained with nth-order Markov model and mapped to one or more orders of Markov models and later used in prediction. Pruning is applied to those examples that are mapped to more than one classifier and by choosing the classifier that predicts accurately with maximum probability. Support Vector Machine (SVM) is used as a prediction technique to create Example Classifier (EC). A testing example is given to EC as an input for appropriate classifier prediction.

Web navigation prediction framework for Web page recommendation is explained in this section. Figure 7.1 represents different stages of this framework. All classifiers are trained on the sessions as training set S and N trained classifiers are derived. In mapping phase, each training session s in S is mapped to one or more classifiers which can be used to predict its target. If any training session s is mapped to more than one classifier, then each trained session is mapped to only one classifier according to the page rank algorithm during filtering process. This PageRank algorithm selects the classifier that predicts correctly with higher incoming page request. If the classifiers have equivalent incoming page request, one classifier is randomly selected. Next, SVM classifier is trained with filtered data. Once SVM classifiers are trained, prediction accuracy and time is calculated on the test dataset. Finally, SVM classifiers are used for page recommendation.

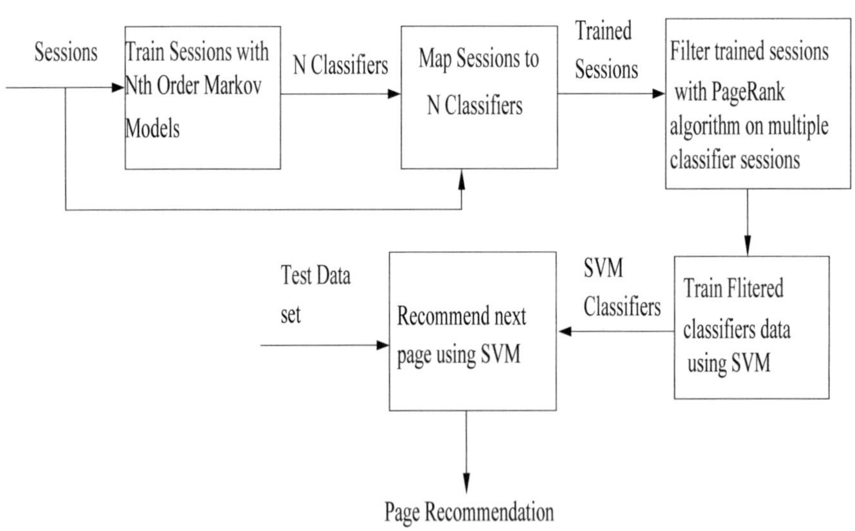

Fig. 7.1 Web navigation prediction framework

7.3.5 *WNPWR Algorithm*

The algorithm of Web Navigation Prediction for Web page Recommendation (WNPWR) is given in Algorithm 7.1.

Algorithm 7.1: WNPWR: Web Navigation Prediction Framework for Web Page Recommendation

---

Input : M is the set of classifier models of size N. S is the set of sessions generated from *session*. T is a test set data.
Output: Recommended Pages

begin
- For each classifier model m in M, train m on S with N-order Markov model.
- For each session s in S and a classifier model m in M
- If m predicts the target s correctly then map s to m.
- For each s in S, if s mapped to more than one classifier then filter it with *PageRank()*.
- Train filtered classifier using SVM.
- For each t in T using SVM classifier
- If classifier predicts t correctly then recommend next page and calculate prediction accuracy.

7.4 Experiments

7.4.1 *Data Collection*

The University of Saskatchewan's (UOFS) and the NASA datasets [29] are used to construct Web navigation prediction framework. The NASA dataset is divided into two groups in order to study the effect of size of the dataset in prediction accuracy and time: $NASA_{Low}$ and $NASA_{Medium}$. In the present work, UOFS data is called $UOFS_{High}$. Log files contain information of the user requests on a particular Web site. The main idea is to analyse HTTP user request and predict Web page access user behaviour. So, all entries with the extension type .gif, .GIF, .jpeg, .JPEG, .jpg are removed. Even status code other than 200, i.e. redirect (300 series), failure (400 series) and state error (500 series) are removed. The statistics of both the dataset is given in Table 7.6.

Each line of HTTP user request contains information about host making the request, timestamp, user request, HTTP reply code and bytes in the reply. If the hostname is not available, the Internet address is considered as host. The timestamp is in the form of *DAY/MON/YYYY: HH:MM:SS -0600*, where *DAY* is the day of the

Table 7.6 Summary of the datasets

	$NASA_{Low}$	$NASA_{Medium}$	$UOFS_{High}$
Total log records	132838	343234	1021891
Number of pages	608	713	3595
Data reduction in % after cleaning	79.09	78.79	58.65

Table 7.7 Sample of HTTP user request

130.54.25.198	01/Jun/1995:00:28:36 -0600	GET/macphed/finite/feresources/node1.html	200	9651
128.171.197.73	01/Jun/1995:00:34:50 -0600	GET/scottp/hawaii	200	29106
130.54.25.198	01/Jun/1995:00:35:01 -0600	GET/macphed/finite/feresources/node59.html	200	2042
sabre45.sasknet.sk.ca	26/Sep/1995:23:48:55 -0600	GET/davs/scn/next.gif HTTP/1.0	200	3003
sabre45.sasknet.sk.ca	26/Sep/1995:23:49:04 -0600	GET/davs/scn/scn3.htm HTTP/1.0	200	1136

month, *MON* is the name of the month, *YYYY* is the year, *HH:MM:SS* is the time of day using a 24-h clock, the timezone is −0600. Table 7.7 shows the sample of the HTTP user request.

7.4.2 User and Session Identification

In HTTP user request, the user is identified by either direct hostname or IP address. As discussed in background, pages accessed by users are divided into distinct session through a time-out in session identification. Sessions are identified by two methods described in background: Method 1, where the time between pages requests exceeds 30 min [26] and Method 2, where the average time of visiting Web pages [27] is computed. The statistics of session identification with Method 1 is given in Table 7.8 and with Method 2 is given in Table 7.9. Sessions are identified by calculating the average time of visiting Web pages for our Web navigation framework as Method 1 has a few disadvantages as discussed earlier.

Table 7.8 Statistics of session identification with traditional method (30 min time-out)

	$NASA_{Low}$	$NASA_{Medium}$	$UOFS_{High}$
Total sessions	5117	13298	66886
Average session length	4.55	4.60	5.30

Table 7.9 Statistics of session identification with average time of visiting Web pages

	$NASA_{Low}$	$NASA_{Medium}$	$UOFS_{High}$
Total sessions	5392	13890	70016
Average session length	3.97	4.02	4.37

7.4.3 Experimental Setup

The setup of Web Navigation Prediction Framework (WNPFramework) is as follows. Sessions are created by calculating the average time on visiting Web pages. Fourth-order Markov models are generated by employing sliding windows on the session set. Four classifiers are represented with these prediction models, namely first-, second-, third- and fourth-order Markov models. Each session is then mapped to one or more orders of Markov models. If any session is mapped to more than one Markov model, then pruning/filtering process is initiated to map to only one classifier. Filtering process is conducted based on the PageRank algorithm, i.e. by selecting the classifier which accurately predicts with the maximum incoming page request. If classifiers have equivalent incoming page request, one classifier is selected randomly. SVM Classifiers are generated by applying training on the filtered data. LIBSVM package [30] is used for SVM implementation. C-SVC and RBF kernel are used as values of the parameters *svm_type* and *kernel_type*, respectively. For other parameters, default values are used.

Prediction accuracy and time is computed to suggest page recommendation. For prediction, given a session s of length L, prediction is conducted using *(L-1)* gram Markov model. The last page of s is concluded to measure the accuracy of the model. If session s is longer than fourth-order Markov model (as fourth-order Markov model is used in our experiments) then sliding window of size four is applied. For example, suppose $s = A, B, C, D, E, F$ then break s into A, B, C, D, E and B, C, D, E, F. Once SVM classifiers are trained and prediction setup is set, prediction accuracy and time are calculated on test dataset. Finally, SVM classifiers can be used for page recommendation. From this instance, setup is called as *WNPFramework + AvgtimeSession*.

The setup of Two-tier Prediction Framework (*TPFramework*) is as follows. Sessions are created by setting 30 min as a default time-out and by calculating the average time on visiting Web pages. Fourth-order Markov models are generated on both session data. Each session is mapped to one or more orders of Markov models and pruning process is applied by selecting the classifier which accurately predicts with the maximum probability. SVM classifiers are generated on the pruned dataset. Once SVM classifiers are trained, prediction time and accuracy is measured (*TPFramework + TraditionalSession*).

In order to study the effect of sessions created by computing the average time on visiting Web pages, prediction accuracy and time is calculated by two-tier prediction framework again but sessions are created by calculating average time on visiting Web pages and are used as training dataset (*TPFramework* + *AvgtimeSession*).

7.4.4 Results Comparison

In this section, the results of our experiments are presented and discussed. Web Navigation Prediction Framework (*WNPFramework*) and Two-tier Prediction Framework (*TPFramework*) are studied and compared. Experiments have been conducted on 4GB memory and Intel(R) Core(TM)2 Duo P7450@2.13 GHz processor. Dataset used for both *WNPFramework* and *TPFramework* are same as discussed in data collection.

Figures 7.2, 7.3 and 7.4 present a comparison between *TPFramework* and *WNPFramework* in terms of prediction accuracy by using different training dataset percentages on all the three datasets. The average accuracy for *WNPFramework* is increased to 44.07, 11.10 and 4.51% for $NASA_{Low}$, $NASA_{Medium}$ and $UOFS_{High}$ datasets, respectively, in comparison to *TPFramework*. It is observed from Fig. 7.3 that accuracy increases for both the frameworks with the increase in the size of the training dataset. But for $NASA_{Medium}$ dataset, at 50 and 60% training dataset, accuracy decreases for *TPFramework*. In *TPFramework*, sessions are created with the traditional method with 30 min time-out. It is observed in $NASA_{Medium}$ dataset that the actual user sessions are more than 30 min, hence these sessions have misclassified classifiers and accuracy is decreased. In *WNPFramework*, sessions are created

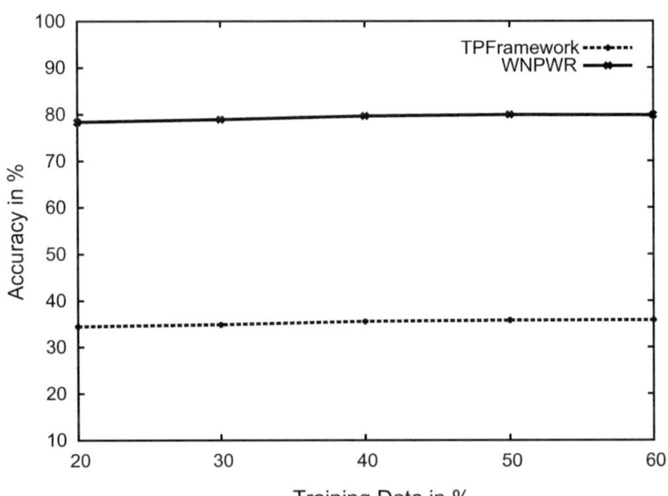

Fig. 7.2 Prediction accuracy comparison between TPFramework and WNPFramework for $NASA_{Low}$

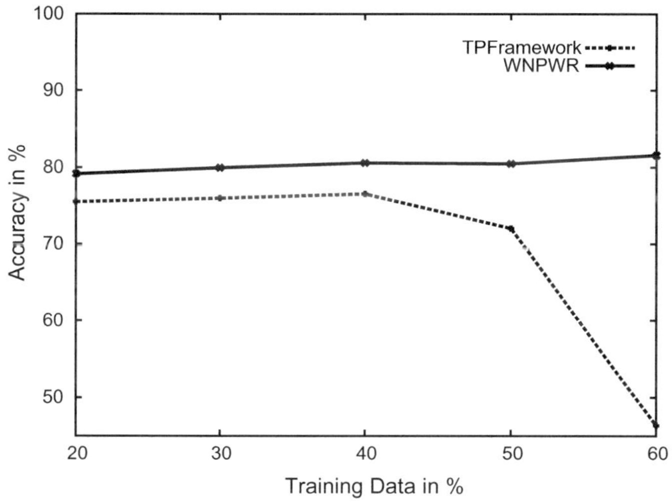

Fig. 7.3 Prediction accuracy comparison between TPFramework and WNPFramework for $NASA_{Medium}$

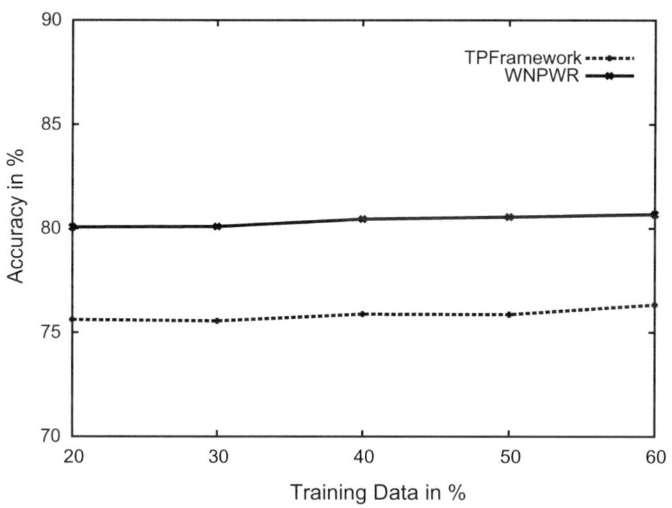

Fig. 7.4 Prediction accuracy comparison between TPFramework and WNPFramework for $UOFS_{High}$

by calculating the average time of visiting Web pages, hence accuracy increases with an increase in training datasets.

Figures 7.5, 7.6 and 7.7 present a comparison between *TPFramework* and *WNPFramework* in terms of prediction time by using different training dataset percentages on all the three datasets. It is observed from Figs. 7.5, 7.6 and 7.7 that

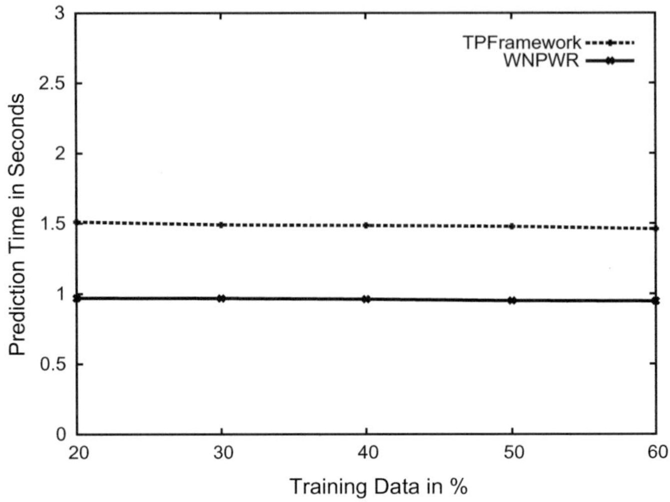

Fig. 7.5 Prediction time comparison between TPFramework and WNPFramework for $NASA_{Low}$

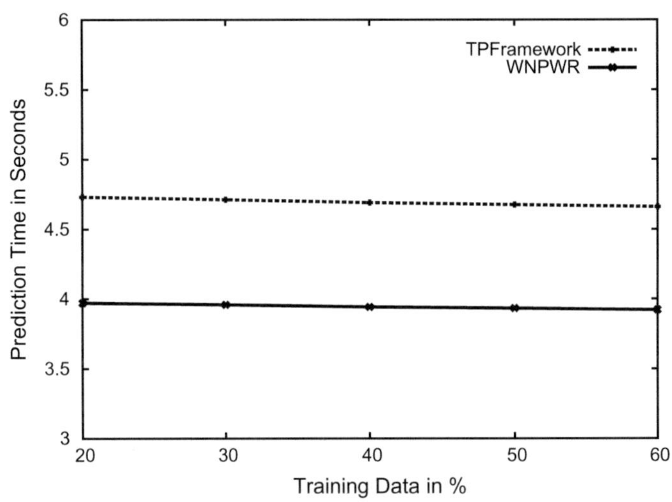

Fig. 7.6 Prediction time comparison between TPFramework and WNPFramework for $NASA_{Medium}$

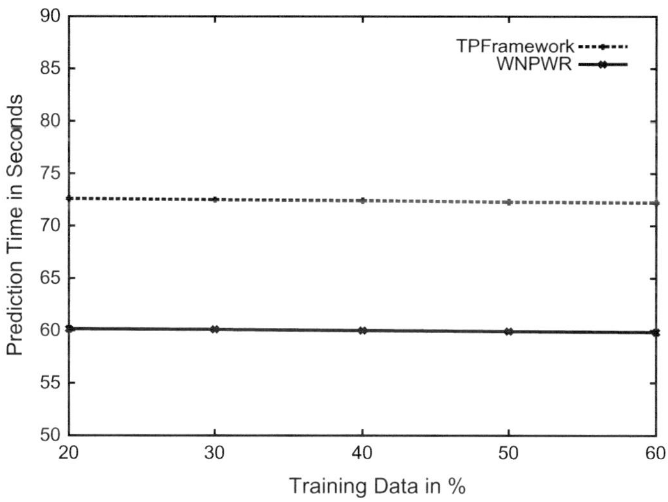

Fig. 7.7 Prediction time comparison between TPFramework and WNPFramework for $UOFS_{High}$

Table 7.10 Prediction accuracy comparison of TPFramework for $NASA_{Low}$, $NASA_{Medium}$ and $UOFS_{High}$ datasets with traditional and average time sessions

		Training data in %				
		20	30	40	50	60
$NASA_{Low}$	TPFramework [15] + Traditional session	1.511	1.490	1.485	1.478	1.460
	TPFramework [15] + Average time session	0.982	0.971	0.965	0.955	0.950
$NASA_{Medium}$	TPFramework [15] + Traditional session	4.731	4.712	4.690	4.675	4.660
	TPFramework [15] + Average time session	3.982	3.970	3.962	3.951	3.941
$UOFS_{High}$	TPFramework [15] + Traditional session	72.591	72.474	72.391	72.265	72.198
	TPFramework [15] + Average time session	60.888	60.781	60.677	60.591	60.481

prediction time decreases with increase in training datasets and decrease in testing datasets. The average prediction time for *WNPFramework* is decreased to 35.35, 15.94 and 17.11% for $NASA_{Low}$, $NASA_{Medium}$ and $UOFS_{High}$ datasets, respectively, in comparison to *TPFramework*.

Table 7.10 presents prediction accuracy comparison between *TPFramework* and *WNPFramework* for $NASA_{Low}$, $NASA_{Medium}$ and $UOFS_{High}$ datasets with traditional and average time sessions. It is observed from Table 7.10 that the average

Table 7.11 Prediction time in seconds comparison of TPFramework for $NASA_{Low}$, $NASA_{Medium}$ and $UOFS_{High}$ datasets with traditional and average time sessions

		Training data in %				
		20	30	40	50	60
$NASA_{Low}$	TPFramework [15] + Traditional session	34.431	34.838	35.520	35.774	35.838
	TPFramework [15] + Average time session	78.366	78.899	79.658	79.919	80.217
$NASA_{Medium}$	TPFramework [15] + Traditional session	75.483	75.943	76.520	72.025	46.367
	TPFramework [15] + Average time session	79.145	79.950	80.587	80.478	81.578
$UOFS_{High}$	TPFramework [15] + Traditional session	75.613	75.557	75.883	75.863	76.338
	TPFramework [15] + Average time session	80.061	80.094	80.457	80.552	80.681

accuracy is increased to 44.12, 11.08 and 4.51% for $NASA_{Low}$, $NASA_{Medium}$ and $UOFS_{High}$ datasets, respectively, for *TPFramework + AvgtimeSession* in comparison to *TPFramework + TraditionalSession.*

Table 7.11 presents prediction time (in seconds) comparison between *TPFramework* and *WNPFramework* for $NASA_{Low}$, $NASA_{Medium}$ and $UOFS_{High}$ datasets with traditional and average time sessions. It is observed from Table 7.11 that the average prediction time is decreased to 35.03, 15.60 and 16.16% for $NASA_{Low}$, $NASA_{Medium}$ and $UOFS_{High}$ datasets, respectively, for *TPFramework + AvgtimeSession* in comparison to *TPFramework + TraditionalSession.*

The major difference between our proposed framework and two-tier framework is the methodology of session identification (in our framework Method 2 which is discussed in user and session identification section). From Table 7.9, it can be observed that Method 2 has generated more number of sessions compared to Method 1 which is used as training dataset. In Method 2, the average visiting time depends on the visiting Web pages, so sessions are mapped to a realistic value than when a single constant value is used. *TPFramework* and *WNPFramework* with AvgtimeSession results in high prediction accuracy on account of these reasons. Even the average session length is small in Method 2 compared to Method 1, that results is lower prediction time. It is also observed that session generation time with Method 2 is more compared to Method 1.

In *WNPFramework*, if any session is mapped to more than one Markov models, then filtering is used by selecting the classifier which accurately predicts with the maximum incoming page request. In the two-tier framework, filtering is used by selecting the classifier which accurately predicts with the maximum probability. For example, user *X* has visited different Web pages in different sessions as shown in Table 7.4. For the test session *(A, B, C, D)*, two classifiers *E* and *F* are suggested.

During filtering E is selected according to the highest probability of classifier, while F is selected as there is more traffic towards F according to the highest incoming page request. *TPFramework* and *WNPFramework* have been compared with AvgtimeSession and it is observed that there is not much difference in prediction accuracy, but the overall prediction time is lower in *WNPFramework* in comparison to *TPFramework*.

7.5 Summary

In this chapter, we have proposed Web Navigation Prediction Framework for Web page Recommendation which creates and generates a classifier based on sessions as training examples. Sessions are created by calculating the average time on visiting Web pages, which maps to the realistic better value than when a single constant value is used. Each session is mapped to one or more generated classifiers and filtering process is applied by the PageRank algorithm. Simulations are performed on UOFS and NASA dataset and are compared with two-tier prediction framework. The WNPWR algorithm outperforms two-tier prediction framework by providing high prediction accuracy with reduced prediction time. Further, it is planned to extend this chapter for online Web page recommendations for mobile applications.

References

1. S. Chimphlee, W. Chimphlee, N. Salim, M.S.B. Ngadiman, Using hybrid markov model for web access prediction. J. Inf. Technol. **3**(3), 86–91 (2012)
2. J. Borges, M. Levene, Evaluating variable-length markov chain models for analysis of user web navigation sessions. IEEE Trans. Knowl. Data Eng. **19**(4), 441–452 (2007)
3. M. Deshpande, G. Karypis, Selective markov models for predicting web-page accesses. ACM Trans. Internet Technol. **4**(2), 163–184 (2004)
4. A. Guerbas, O. Addam, O. Zaarour, M. Nagi, A. Elhajj, M. Ridley, R. Alhajj, Effective web log mining and online navigational pattern prediction. J. Knowl. Based Syst. **49**(2), 50–62 (2013)
5. C. Dimopoulos, C. Makris, Y. Panagis, E. Theodoridis, A. Tsakalidis, A web page usage prediction scheme using sequence indexing and clustering techniques. J. Data Knowl. Eng. **69**(4), 371–382 April (2010)
6. C.-H. Lee, Y.-L. Lo, Y.-H. Fu, A novel prediction model based on hierarchical characteristics of web site. Int. J. Expert. Syst. Appl. **38**(4), 3422–3430 (2011)
7. K. Dembczyński, W. Kotłowski, M. Sydow, Effective prediction of web user behavior with user-level models. J. Fundam. Inform. **89**(3), 189–206 (2008)
8. C. Liu, R.W. White, S. Dumais, Understanding web browsing behaviors through Weibull analysis of dwell time, in *The Proceedings of 33rd Annual International ACM SIGIR Conference on Research and Development in Information Retrieval, SIGIR '10* (2010), pp. 379–386
9. F.K.H. Phoa, J. Sanchez, Modelling the browsing behavior of world wide web users. Open J. Stat. **3**(2), 145–154 (2013)
10. R.W. White, P. Bailey, L. Chen, Predicting user interests from contextual information, in *The Proceedings of International ACM SIGIR Conference on Research and Development in Information Retrieval* (2009), pp. 363–370

11. P. Thwe, Proposed approach for web page access prediction using popularity and similarity based pagerank algorithm. Int. J. Sci. Technol. Res. **2**(3), 240–246 (2013)
12. F. Khali, J. Li, H. Wang, Integrating recommendation models for improved web page pediction accuracy, in *The Proceedings of the 31st Australasian Conference on Computer Science, ACSC '08* (2008), pp. 91–100
13. M.A. Awad, L.R. Khan, Web navigation prediction using multiple evidence combination and domain knowledge. IEEE Trans. Syst., Man Cybern.-Part A: Syst. Hum. **37**(6), 1054–1062 (2007)
14. R. Dutta, A. Kundu, D. Mukhopadhyay, Clustering-based web page prediction. Int. J. Knowl. Web Intell. **2**(4), 257–271 (2011)
15. M.A. Awad, I. Khalil, Prediction of user's web-browsing behavior: application of markov model. IEEE Trans. Syst., Man Cybern.-Part B: Cybern. 42(4), 1131–1142 (2012)
16. V.S. Tseng, K.W. Lin, Efficient mining and prediction of user behavior patterns in mobile web systems. J. Inf. Softw. Technol. **48**(6), 357–369 (2006)
17. G. Zhao, W. Lai, Predicting user behavior in mobile internet based on random walk. J. Comput. Inf. Syst. **9**(22), 9157–9164 (2013)
18. C.-M. Huang, J.J.-C. Ying, V.S. Tseng, Mining users' behaviors and environments for semantic place prediction, in *Mobile Data Challenge Workshop* (2012)
19. M. Silic, G. Delac, I. Krka, S. Srbljic, Scalable and accurate prediction of availability of atomic web services. IEEE Trans. Serv. Comput. **7**(2), 252–264 (2014)
20. J. Huang, R.W. White, Parallel browsing behavior on the web, in *The Proceedings of the 21st ACM Conference on Hypertext and Hypermedia, HT'10* (2010), pp. 13–18
21. S. Goel, J.M. Hofman, M.I. Sirer, Who does what on the web: a large-scale study of browsing behavior, in *The Proceedings of 6th AAAI International Conference on Weblogs and Social Media, AAAI' 12* (2012), pp. 130–137
22. Z. Cheng, B. Gao, T.-Y. Liu, Actively predicting diverse search intent from user browsing behaviors, in *The Proceedings of 19th International Conference on World Wide Web, WWW '10* (2010), pp. 221–230
23. Y. Zhang, W. Chen, D. Wang, Q. Yang, User-click modeling for understanding and predicting search-behavior, in *The Proceedings of 17th ACM SIGKDD International Conference on Knowledge Discovery and Data Mining, KDD '11* (2011), pp. 1388–1396
24. X. Tian, Y. Lu, L. Yang, Query difficulty prediction for web image search. IEEE Trans. Multimed. **14**(4), 951–962 (2012)
25. J. Yu, Y. Rui, D. Tao, Click prediction for web image reranking using multimodal sparse coding. IEEE Trans. Image Process. **23**(5), 2019–2032 (2014)
26. R. Cooley, B. Mobasher, J. Srivastava, Data preparation for mining world wide web browsing patterns. J. Knowl. Inf. Syst. **1**(1), 5–32 (1999)
27. C.E. Dinuca, D. Ciobanu, Improving the session identification using the mean time. Int. J. Math. Model. Methods Appl. Sci. **6**(2), 265–272 (2012)
28. S. Brin, L. Page, The anatomy of a large-scale hypertextual web search engine. J. Comput. Netw. **56**(18), 3825–3833 (2012)
29. Internet Traffic Archive, http://ita.ee.lbl.gov/html/contrib/Sask-HTTP.html
30. C.-C. Chang, C.-J. Lin, LIBSVM: a library for support vector machines. ACM Trans. Intell. Syst. Technol. **2**(3), 1–39 (2011)

Chapter 8
Web Page Recommendations Based on User Session Graph

K. R. Venugopal and Sejal Santosh Nimbhorkar

Abstract Web page recommender system provides users with recommendations to assist their navigation and increases the website usability and user satisfaction. In this chapter, Web page recommendation method is presented by constructing User Session Graph using user sessions from the navigation log. The node represents Web pages and weight on the edge is calculated by the number of times the Web pages present in the sessions. *new page problem* is solved by computing co-occurrence value between two terms present in titles. Web pages are recommended based on connected nodes in the graph and co-occurred terms. Experiments are conducted on user navigation data collected from Microsoft Website www.microsoft.com. The proposed method is compared with *TermNetWP* method and outperforms *TermNetWP* with higher precision and satisfaction values.

8.1 Introduction

Internet usage has increased excessively as a result of evolution in e-commerce, research, e-banking, education, news, music, movies and electronics devices. Hence, a huge amount of information is archived and it keeps growing rapidly without any control. The decreasing costs in secondary storage have made it easy to store a huge volume of information. The valuable information is beneficial to determine interesting and useful patterns that are used by many researchers for guiding the users to visit the Web pages during their activity on the Web. This type of system is called Recommender System and helps to predict the user request.

The Web page recommender system provides users with recommendations to assist their navigation. The objective is to discover which Web page user will access next. Web page recommendation increases the website usability and user

K. R. Venugopal (✉)
Bangalore University, Jnana Bharathi, Bengaluru 560056, India
e-mail: venugopalkr@gmail.com

S. Santosh Nimbhorkar
BNM Institute of Technology, Banashankari, Bengaluru 560070, India
e-mail: sej_nim@yahoo.co.in

K. R. Venugopal et al., *Web Recommendations Systems*,
https://doi.org/10.1007/978-981-15-2513-1_8

satisfaction. It is also useful in many applications like Web caching, prefetching and online advertising.

There are many problems in generating efficient Web page recommendation framework such as extracting useful information of the domain, efficiently determine knowledge and navigational patterns from available user navigational data, employing extracted knowledge to build a model and develop powerful Web page recommendation system.

Researchers have developed various methods to solve the above problems using available Web usage data. Probabilistic models and tree construction from Web log can effectively produce useful patterns [1–4] and are trained from navigational data to build links between Web pages. For a given Web page sequence as input to these trained models on previously visited pages, the Web pages user will access next can be predicted. These approaches requires more training dataset to get higher prediction accuracy and solely dependent on the Web usage data. Hence, if the input Web page is not present in usage data then it fails to predict the Web pages and is called *new page problem*.

The *new page problem* can be solved by incorporating semantic relation [5] and domain knowledge [6–8]. A semantic of Web pages of a website is represented by the domain ontology and combined with usage data to improve the Web page recommendation performance. Combining semantic knowledge to Web usage data has gained greater performance than traditional Web usage mining algorithms. However, the construction of efficient domain ontology and representation of semantic domain knowledge is a big challenge in these models.

In this chapter, Web page framework is presented where user session graph is constructed from sessions collected from user navigation log. In this graph, the nodes represent Web pages and an edge exists between two nodes if two Web pages are present in the session. The weight on the edge is calculated by the number of times the Web pages are present in sessions. This graph shows the relation between the two visited Web pages by the users. For a given Web Access Sequence (WAS) the connected Web pages with its weight are extracted for each page present in WAS. If any *new page* exists in WAS, then the Web pages are extracted based on co-occurred term calculation with its weight. Extracted Web pages are arranged in the descending order based on its weight and Top-*n* Web pages are recommended.

8.2 Related Works

In this section, various webpage recommendation techniques are discussed. Murat et al. [9] presented hybrid recommendation systems that combine existing recommendation methods based on Web usage. The comparative evaluation is presented by combining k-means models, Markov models, Association rule mining model and Click Stream Tree model. It shows the combination of various methods that affect the prediction accuracy.

Usually, in a website hierarchical link is created to organize webpages without any semantic relation. Hence, the traditional link-based algorithm for Web page retrieval gives poor performance. Li et al. [10] have proposed a Hierarchical Navigation Path based method for Web page retrieval. In this method, Web textual information such as page titles, anchor text and URLs are incorporated with website link structure for Web page retrieval. This method improves Web page retrieval accuracy compared to traditional link-based algorithm. However, the performance decreases with increase in path length. Markov models with semantic information is used for Web perfecting [11]. Semantic information is incorporated into Markov transition matrix. Hence, it can provide better prediction and solve the problem of contradicting prediction. This semantic information is used to prune states in Selective Markov models.

A technique that combines domain knowledge and Web navigation log is proposed [12] for prediction. It integrates the conceptual hierarchy of website and usage information based on biological sequence alignment. Similarity score between pages is computed based on conceptual hierarchy. Time spent on page and browsing order is incorporated in similarity calculation. A method that combines navigational patterns and page connectivity is proposed for Web page recommendation [2]. User concepts from Web log is used to extract navigational patterns and used to produce Web page recommendations by accomplishing navigational behaviour of a user. Recommended Web pages are ranked by computing hub score from page connectivity knowledge.

A hybrid method is presented that minimizes repetitive Web surfing for personalized new recommendation [13]. Web pages of Web navigational log are classified by computing weight of terms and used to construct user's interest model. User's preference model is constructed from hyperlink of user navigational log. The recommendation probability for a user is computed by matching user's models and news content. Forsati and Meybodi [14] have proposed a hybrid algorithm based on weighted association rule and distributed learning automata for Web page recommendation. Users' navigational pattern is extracted by association rule mining and used for recommendation based on user's current status. In distributed learning automata, link analysis and behaviour of users are integrated to assign probability to the webpages. Recommendations are made based on structure of website and user behaviour patterns. HITS algorithm is used to extend the recommendation set.

A method that integrates prefetching and caching is presented to derive prefetching rules [15]. The cyclic behaviour and periodicity of Web access sequence is used to fetch profit-driven caching policy. The experiment results show that hit ratio is increased for the cached objects. The Probability of Correctness of Facts is proposed to compute Trustworthiness of websites to rank the Web pages [16]. Probability-based similarity function is used to extract correctness by matching facts provided by Web pages and Web pages' facts.

Collaborative social annotations method is applied for Web page recommendations [17]. Semantic clusters are generated from users' information and Web pages. Social annotation clusters are also generated from annotation and Web pages. Semantic and social annotation clusters are used to recommend Web pages. Collaborative filtering and Topic-aware Markov Model is used for personalized Web page recommendation [18]. User's navigational patterns is learned from topic aware Markov model. It acquires topical relevance of pages. Collaborative filtering is used to find user similarities and integrated to rank the recommended webpages.

8.3 Web Page Recommendations Framework and WRUSG Algorithm

8.3.1 Problem Definition

Given a Web Access Sequence (WAS) from user navigation *log*, the objective is to recommend webpages. It is assumed that the user is online and the user navigation *log* is available.

8.3.2 Web Page Recommendations Framework

In this section, Web Page Recommendation Framework is presented that solves new page problem. User session graph is constructed from the user navigation history in which nodes represent Web pages and edge represent navigation between the Web pages. For a given Web page sequence as an input, Web pages are recommended by travelling user session graph. For the new page problem, co-occurrence of domain terms is computed and later used to recommend Web pages. Here, first few terms are explained.

Web Pages: The domain consists of numerous pages with meaningful information. Each page has small description about it as a title of that page. A domain can be represented as $D = \{P_1, P_2,\ldots,P_m\}$ with m number of pages.

Domain Terms: A title consist of set of definite words which gives the knowledge of that page. Domain terms are the words present in the Web pages' titles for a particular domain and represents as $T = \{t_1, t_2,\ldots,t_n\}$.

Web Access Sequence (WAS): The user navigation log consists of user *id* and the sequence of Web pages a user has visited during browsing and it is called Web access sequence. This framework has the following steps:

Step 1: User Session Graph Construction

User session graph is constructed based on Web navigation history that gives us the knowledge about the users' Web access sequence. This sequence contains user clicked patterns, transactions or user interactions with Web resources. A session is the continuous Web access sequence visited by a user in a particular time period. It gives information about the visited pages and its association between Web pages. The visited Web page in a particular session can also be present in other sessions. The individual session cannot infer any kind of relationship between different Web pages. Hence, user session graph is constructed by combing all the user sessions.

Consider a Web navigation log with m number of sessions $S_m = \{S_1, S_2,\ldots,S_m\}$. Each session consists of Web pages visited by users $S = \{P_1, P_2,\ldots,P_n\}$. A User Session directed Graph $USG_{pp} = (V_{pp}, E_{pp})$ is constructed, where V_{pp} is the set of visited Web pages in the sessions S_m and E_{pp} is the set of edges between two Web pages. The edge (P_i, P_j) exist if and only if Web page P_i follows Web page P_j in a session S. The weight W_{pp} on the edge is calculated by number of times (P_i, P_j) present in sessions S_m. The weight should be normalized. The Directed user session graph is constructed using Function 8.1.

Consider a Web navigation log with five sessions $L = \{S_1, S_2, S_3, S_3, S_4\}$. The Web Access Sequence (WAS) in each session is shown in Table 8.1. Figure 8.1 shows the User Session directed Graph (USG) constructed from data present in Table 8.1. It is observed from the sessions that users have visited seven unique webpage during their navigation, hence the USG has seven nodes. In WAS, user has visited webpage P_1 to P_2 two times, hence the weight on that edge is two. Similarly, the weight on P_2 to P_4 is 2, P_4 to P_5 is 4 and so on.

Step 2 : Compute Co-occurrence of Domain Terms

The user session graph is constructed based on user navigation history. If a Web page of a domain is not visited by a user then that page does not appear in the graph. Hence, Web pages cannot be recommended for unvisited pages. Relation between the domain terms plays an important role to overcome this problem. It acts as a bridge between the current page and the page to be recommended. Co-occurrence of domain terms revels relations between the terms explicitly. Higher the Co-occurrence value, more relation between the domain terms. This co-occurrence value is helpful to find relation between two Web pages based on the terms present in the title of that page. The co-occurrence between two terms t_i and t_j is represented as $C(t_i, t_j)$ and its value is set to 1 if t_i and t_j occurs together in the session. Further, it increments when they appear together in another session. The value of $C(t_i, t_j)$ is normalized by using Eq. 8.1.

$$NormalizedC(t_i, t_j) = CN(t_i, t_j) = \frac{C(t_i, t_j)}{C(t_i)} \tag{8.1}$$

Here, $C(t_i)$ is the number of times the term t_i that occurs in all the sessions.

Step 3: Extract Web Pages for new page problem using Co-occurrence Value

If any *new page* is present in the input sequence, then the domain terms D_i of that page are considered. The co-occurrence values of each term in D_i are computed as explained in step 2. Arrange the co-occurred terms according to their co-occurrence value in the descending order. All the Web pages that contain co-occurred term with the co-occurrence value as the weight $W_{co-occur}$ are extracted. Web pages are extracted for the *new page* using Function 8.2.

Function 8.1: Directed User Session Graph Generation

Function: GraphConstuction
Data: Consider User Navigation Log l, Generate Directed User Session Graph

Extract Web Access Sequence from Log l
For each user extract session from WAS
Let $Session_{count}$ = Total sessions in WAS
Let Directed User Session Graph $USG_{pp} = (V_{pp}, E_{pp})$ = NULL
// USG construction from WAS
for *i = 1 to* $Session_{count}$ **do**
  for *each* $session_i$ **do**
    Let $Page_{count}$ = Number of visited pages in $Session_i$
    for *j = 1 to* $Page_{count}$ **do**
      if $Page_j$ *and* $Page_{j+1}$ *is in* V_{pp} **then**
        Increment W_{pp} of $(Page_j, Page_{j+1})$
      else if $Page_j$ *is present in* V_{pp} *and* $Page_{j+1}$ *is not present* **then**
        Add $Page_{j+1}$ to V_{pp}
        Add edge between $Page_j$ and $Page_{j+1}$ to E_{pp}
        Set W_{pp} of $(Page_j, Page_{j+1}) = 1$
      else if $Page_{j+1}$ *is present in* V_{pp} *and* $Page_j$ *is not present* **then**
        Add $Page_j$ to V_{pp}
        Add edge between $Page_j$ and $Page_{j+1}$ to E_{pp}
        Set W_{pp} of $(Page_j, Page_{j+1}) = 1$
      else
        Add $Page_j$ and $Page_{j+1}$ to V_{pp}
        Add edge between $Page_j$ and $Page_{j+1}$ to E_{pp}
        Set W_{pp} of $(Page_j, Page_{j+1}) = 1$

// Normalize weight of USG
Let $Node_{count}$ = Number of nodes in V_{pp}
for *m = 1 to* $node_{count}$ **do**
  Let W_{node_m} = Number of times $node_m$ is present in WAS
  for *each connected node* C_{node} *from* $node_m$ **do**

$$NormalizeW(node_m, C_{node}) = \frac{W_{pp}(node_m, C_{node})}{W_{node_m}}$$

Table 8.1 Web access sequence in session

Session	Web access sequence
S_1	P_1, P_2, P_4, P_5, P_9
S_2	P_1, P_2, P_5
S_3	P_3, P_4, P_5, P_7
S_4	P_2, P_4, P_5
S_5	P_4, P_5, P_9

Fig. 8.1 User Session Graph

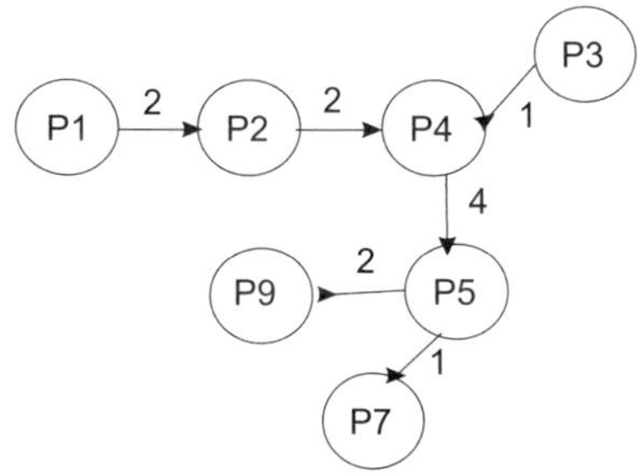

Function 8.2: Web Pages Extraction for *new page*

Function: NewPage
Data: Consider *new page* as Input and Web pages with its weight are extracted

Consider Domain terms $D = \{t_1, t_2, \ldots, t_n\}$ of $page_{newpage}$
for *each term* t_i *in D* **do**
    Co-occurred Terms C_{terms}[ ][ ] = Co-occurred terms, co-occurrence value
Arrange C_{terms}[ ][ ] in descending order of co-occurrence value
Extract web-pages which are containing Co-occurred terms
Return extracted web-pages and their corresponding Co-occurred terms' Weight as $W_{co-occur}$

Step 4: Web Page Recommendation

In this step, the strategy is defined for recommending the Web pages. User session graph is constructed with training sessions as explained in step 1. From the testing sessions, Web access sequence is given as an input to the graph and for each page in the sequence, Web pages connected to that page with the weight W_{pp} are considered. If any *new page* is present in the input sequence then the Web pages are extracted as explained in step 3 with their weight $W_{co-occur}$. The resultant Web pages are arranged in the descending order based on the weight and top-*n* Web pages are recommended.

8.3.3 *WRUSG Algorithm*

The method to recommend the Web pages is shown in Algorithm 8.1.

Algorithm 8.1: WRUSG : Web Page Recommendation using User Session Graph

---

Input : User Navigation Log l
Output: Recommended Web pages

begin
  USG_{pp} = Construct Directed User Session Graph by using Function $GraphConstuction(l)$
  Let $GraphInput[\,]$ = WAS
  for *each page* $Page_{curr}$ *in GraphInput[ ]* **do**
    if $Page_{curr}$ *is present in* USG_{pp} **then**
      $Connected_{node}[\,][\,][\,]$ =$page_{curr}$, connected node of $page_{curr}$ $C_{pagecurr}$, W($page_{curr}$, $C_{pagecurr}$)
    else
      $New_{page}[\,][\,]$ = NewPage($page_{curr}$)
      $Connected_{node}[\,][\,][\,]$ = $page_{curr}$, $New_{page}[\,][\,]$
  Arrange $Connected_{node}[\,][\,][\,]$ in descending order of $W(page_{curr}, C_{pagecurr})$
  Recommend top-*n* pages from $C_{pagecurr}$

8.4 Experiments

8.4.1 *Data Collection*

In this experiment, user navigation data is collected from Microsoft Website www.microsoft.com [19]. This dataset contains 294 unique Web pages and 37711 user sessions. Each Web page has a description of it as a title and each title is a collection of words. The dataset contains 641 total words and 330 unique words in Web pages titles.

8.4.2 *Experimental Setup*

The proposed model is compared with the combination of *TermNetWP* and Conceptual Prediction Model (CPM) presented in [20] and also solves *new page problem*.

The setup of *TermNetWP* + CPM is as follows: Titles of the Web pages are collected and term sequence from titles are extracted. Term frequency of each term is computed and terms are mapped to respective Web pages as per their order of occurrence in title. Co-occurrence value of co occurred words is computed. CPM is computed using co-occurrence value and term frequency of terms. For a given Web Access Sequence (WAS), domain terms are extracted for each page in WAS and co-occurred words are considered for each domain term. Co-occurred terms are arranged in the descending order and Web pages are recommended that contain co-occurred terms. If there are two co-occurred terms are present, then the set-up is called $TermNetWP_1$. If there are three co-occurred terms present, then the setup is called $TermNetWP_2$.

The setup of the proposed method Web Page Recommendations based on User Session Graph (WRUSG) is as follows: The directed user session graph is constructed from user sessions. Co-occurrence value of co-occurred words is computed. For the given Web access sequence, connected Web pages are extracted for each page from the graph with its weight. If any new page occurs in a Web access sequence, then the domain terms are extracted for each page and co-occurred terms are considered for each domain terms. Web pages are extracted that contains co-occurred terms with co-occurrence value as weight. Top-*n* Web pages are recommended based on their weight.

8.4.3 Performance Metrics

The performance metrics used for evaluation are Precision and Satisfaction [21]. The Precision measures the probability that user visits one of the recommended Web pages. The satisfaction measures the probability that user visits next recommended Web pages after visiting one of the recommended pages. Web page recommendation rule needs to be set to compute Precision and Satisfaction.

Web Page Recommendation Rule: Let $WAS = \{p_1, p_2, \ldots, p_m\}$, $m \geq 2$. Web pages are recommended by giving input as $WAS = \{p_1, p_2, \ldots, p_{m-1}\}$ and stored in $Recom = \{rec_1, rec_2, \ldots, rec_n\}$. For any $p_k(1 \leq k \leq m-1)$ (1) If *Recom* contains p_{k+1} then *Recom* is *correct*. (2) If *Recom* contains p_{k+1} to p_m then *Recom* is *satisfied*. (3) If $n = 0$ the *Recom* is *empty*.

Precision and Satisfaction are computed using Eqs. 8.2 and 8.3, respectively.

$$Precision = \frac{| Recom_{correct} |}{| Recom |} \tag{8.2}$$

$$Satisfaction = \frac{| Recom_{satisfie} |}{| Recom |} \tag{8.3}$$

Here, $Recom_{correct}$ is the total number of correct recommendations for given WASs. *Recom* is the total number of recommendations for given WASs. $Recom_{satisfie}$ is the total number of satisfied recommendations for given WASs.

8.4.4 Performance Evaluation

In this section, results are presented and discussed. Performance metrics are used to compare the results of the proposed model and *TermNetWP* + CPM [2]. Experiments have been conducted on 4 GB memory and Intel(R) Core(TM) i5-5200U CPU @ 1.80 GHz processor. The dataset used for evaluation for both the methods are the same as explained in Data collection. Dataset is divided into training and testing dataset. Training and testing dataset contains 32711 and 5000 sessions, respectively. The user session graph is constructed with 32711 sessions in the proposed method. The graph contains 233 nodes and 2448 edges. The minimum support threshold $Threshold_{min}$ is set in order to evaluate the performance of Web access patterns.

Let Web Access Sequences in the testing dataset are $WASs = \{W_1, W_2, \ldots, W_m\}$. Each WAS $W_i = \{p_1, p_2, \ldots, p_n\}$. Threshold of each WAS_i is computed using Eq. 8.4.

$$Threshold_{WAS_i} = \frac{Total_{WAS}}{\mid WASs \mid} \tag{8.4}$$

Here, $Total_{WAS}$ = Number of times WAS_i occurs in $WASs$. The $WASs$ whose $Threshold_{WAS_i}$ is greater than or equal to $Threshold_{min}$ are considered as frequent Web access sequence. If the value of $Threshold_{min}$ is set to smaller value then more number of frequent WAS are found.

Tables 8.2 and 8.3 shows comparison of the precision and satisfaction values for $TermNetWP_1$, $TermNetWP_2$ and the WRUSG. It is observed that the values of precision and satisfaction is higher in WRUSG in comparison with *TermNetWP* models. The precision of the WRUSG is increased by 23.688% and 27.3% in comparison with $TermNetWP_1$ and $TermNetWP_2$ method, respectively. The satisfaction of the WRUSG is increased by 25.736% and 30.816% in comparison with $TermNetWP_1$ and $TermNetWP_2$ method respectively. The WRUSG considered user history first and if any *new page* occurs only semantic knowledge is used to recommend Web pages, while *TermNetWP* combines user history and domain knowledge to recommend Web pages. Hence the WRUSG outperforms the *TermNetWP* method.

8.5 Summary

Web page recommendation method is presented by considering user sessions and co-occurred terms in titles of the Web pages. User session graph is constructed using user sessions from the navigation log in where nodes are Web pages and edges are present

Table 8.2 Comparison of precision values

Minimum support	0.4	0.5	0.6	0.7	1.0
Number of WASs	165	139	120	110	74
$TermNetWP_1$	0.3505	0.3602	0.3839	3725	0.5448
$TermNetWP_2$	0.2720	0.2965	0.3601	0.3856	0.4871
WRUSG	0.6149	0.6397	0.6339	0.6176	0.6602

Table 8.3 Comparison of satisfaction values

Minimum support	0.4	0.5	0.6	0.7	1.0
Number of WASs	165	139	120	110	74
$TermNetWP_1$	0.3084	0.3137	0.3273	0.3104	0.4935
$TermNetWP_2$	0.2203	0.2377	0.2886	0.3169	0.4358
WRUSG	0.5938	0.6127	0.6011	0.5915	0.6410

if those two Web pages exist in sessions. The weight on the edge is computed by the number of times the Web pages are present in sessions. To solve *new page problem*, co-occurrence value between two terms present in titles is computed that is helpful to find relation between the two Web pages. Web pages are recommended based on the connected nodes in the graph and co-occurred terms. Experiments are conducted on the user navigation data collected from Microsoft Website www.microsoft.com. The WRUSG method is compared with $TermNetWP$ method [20] and outperforms $TermNetWP$ by increased precision value of 23.688% and 27.3% and satisfaction value of 25.736% and 30.816% in comparison with $TermNetWP_1$ and $TermNetWP_2$ method, respectively.

References

1. B. Liu, *Web Data Mining: Exploring Hyperlinks, Contents, and Usage Data* (Springer Science & Business Media, New York, 2007)
2. Y.-F, Huang, J.-T. Jhang, Recommendations of personal web pages based on user navigational patterns. Int. J. Mach. Learn. Comput. **4**(4), 307–317 (2014)
3. C.I. Ezeife. Y. Lu, Mining web log sequential patterns with position coded pre-order linked wap-tree. Data Min. Knowl. Discov. **10**(1), 5–38 (2005)
4. J. Borges, M. Levene, Generating dynamic higher-order markov models in web usage mining, in *European Conference on Principles of Data Mining and Knowledge Discovery* (2005) pp. 34–45
5. G. Stumme, A. Hotho, B. Berendt, Usage mining for and on the semantic web, in *National Science Foundation Workshop on Next Generation Data Mining*, (2002) pp. 143–153
6. H. Dai, B. Mobasher, Integrating semantic knowledge with web usage mining for personalization, in *School of Computer Science, Telecommunication, and Information Systems* (2007)

7. S.A. Rios, J.D. Velásquez, Semantic web usage mining by a concept-based approach for off-line web site enhancements, in *IEEE/WIC/ACM International Conference on Web Intelligence and Intelligent Agent Technology, 2008. WI-IAT'08* (2008), pp. 234–241
8. S. Salin, P. Senkul, Using semantic information for web usage mining based recommendation, in *24th International Symposium on Computer and Information Sciences, 2009* (2009), pp. 236–241
9. M. Göksedef, Ş. Gündüz-Öğüdücü, Combination of web page recommender systems. Expert. Syst. Appl. **37**(4), 2911–2922 (2010)
10. J.-Q. Li, Y. Zhao, H. Garcia-Molina, A path-based approach for web page retrieval. World Wide Web **15**(3), 257–283 (2012)
11. N.R. Mabroukeh, C.I. Ezeife, Semantic-rich Markov models for web prefetching, in *2009 IEEE International Conference on Data Mining Workshops* (2009), pp. 465–470
12. A. Bose, K. Beemanapalli, J. Srivastava, S. Sahar, Incorporating concept hierarchies into usage mining based recommendations, in *International Workshop on Knowledge Discovery on the Web* (2006), pp. 110–126
13. H. Wen, L. Fang, L. Guan, A hybrid approach for personalized recommendation of news on the web. Expert. Syst. Appl. **39**(5), 5806–5814 (2012)
14. R. Forsati, M.R. Meybodi, Effective page recommendation algorithms based on distributed learning automata and weighted association rules. Expert. Syst. Appl. **37**(2), 1316–1330 (2010)
15. K.C. Srikantaiah, N.K. Kumar, K.R. Venugopal, L.M. Patnaik, Web caching and prefetching with cyclic model analysis of web object sequences. Int. J. Knowl. Web Intell. 2, **5**(1), 76–103 (2014)
16. K.C. Srikantaiah, P.L. Srikanth, V. Tejaswi, K. Shaila, K.R. Venugopal, Lalit M. Patnaik, Ranking search engine result pages based on trustworthiness of websites. IJCSI Int. J. Comput. Sci. Issues **9**(4), (2012)
17. H.-Q. Li, F. Xia, D. Zeng, F.-Y. Wang, W.-J. Mao, Exploring social annotations with the application to web page recommendation. J. Comput. Sci. Technol. **24**(6), 1028–1034 (2009)
18. Q. Yang, J. Fan, J. Wang, L. Zhou, Personalizing web page recommendation via collaborative filtering and topic-aware markov model, in *2010 IEEE International Conference on Data Mining* (2010), pp. 1145–1150
19. http://kdd.ics.uci.edu/databases/msweb/msweb.html
20. T.T.S. Nguyen, H.Y. Lu, J. Lu, Web-page recommendation based on web usage and domain knowledge. IEEE Trans. Knowl. Data Eng. **26**(10), 2574–2587 (2014)
21. B. Zhou, S.C. Hui, A.C.M. Fong, Efficient sequential access pattern mining for web recommendations. Int. J. Knowl.-Based Intell. Eng. Syst. **10**(2), 155–168 (2006)

Chapter 9
Advertisement Recommendations Using Expectation Maximization

K. R. Venugopal and Sejal Santosh Nimbhorkar

Abstract The advertiser has to understand the purchase requirement of the users who are looking for a particular service to recommend advertisement. Once the users' demand is identified, advertisers can target those users with an appropriate query. In this chapter, predicting conversion in advertising using expectation–maximization [PCAEM] model is proposed to provide an influence of their advertising campaigns to the advertisers by understanding hidden topics in search terms with respect to the time period. Query terms present in search log are used to construct vocabulary. Expectation–Maximization technique is used to learn hidden topics from the vocabulary. Least Absolute Shrinkage and Selection Operator (LASSO) is used to predict total number of conversions. Experiment results show that PCAEM model outperforms TopicMachine model by reducing Root Mean Squared Error (RMSE) and Mean Absolute Error (MAE) for prediction.

9.1 Introduction

Internet provides information to users; it takes input as a query and gives related results. Many service providers use these characteristics of the Internet to their advantage. Consider a query *create an android app for generating recipe* is entered by a user to the search engine. The user may be looking for online tutorials or classes for learning Android programming. It is assumed that such a user is willing to pay for a service that can help him to learn Android program. The service provider that provides such services wants their advertisement to appear as the most relevant result for this query.

The query given to a search engine is called as the search term and resulting advertisements are called as keyword-based advertising. The search engine market

K. R. Venugopal (✉)
Bangalore University, Jnana Bharathi, Bengaluru 560056, India
e-mail: venugopalkr@gmail.com

S. Santosh Nimbhorkar
BNM Institute of Technology, Banashankari, Bengaluru 560070, India
e-mail: sej_nim@yahoo.co.in

K. R. Venugopal et al., *Web Recommendations Systems*,
https://doi.org/10.1007/978-981-15-2513-1_9

(SEM) agencies connect the two sides of this same coin, i.e. the customers and the service providers. The SEM agencies maintain a large collection of keywords for their clients, the service providers. As the services provided by these clients grow, the number of keywords also increases and needs to be managed. SEM agencies conduct experiments on the keywords to test their weight, relevance and usage. These tests help the clients to select the best variant for the keywords related to the services they provide. However, with the increase in the number of keywords and many variants to consider, the advertisement management becomes a burden on even the most experienced advertisers.

The advertiser needs to understand the underlying requirements of the users from the queries. These requirements are usually the descriptions of service. From the query, *create an android app for generating recipe*, the user might actually require *I want to create an android application where I can give picture of the ingredients that are available to me as an input and the generated application gives the relevant recipes out of these ingredients*. The result of the query must be relevant based on the information need present in it and not because it contains the keywords of the description.

Advertisers have to analyse the purchase requirement of the targeted users that helps them to aim those users with appropriate search terms. Advertises analyse query logs, search keywords reports and trend reports to determine relevant search keywords. It is very difficult to understand the latent need of the user from a few words in the query and how an advertiser characterizes an information need [1, 2]. Hence, it is necessary to develop a model on top of the search campaigns that the advertisers can examine the consequence of the topical changes in trends over the time and target new market.

In this chapter, predicting conversion in advertising using expectation–maximization [PCAEM] model is proposed to provide the effectiveness of their advertising campaigns to the advertisers with respect to time period. Vocabulary is constructed from the query keywords present in search query log. Gaussians are assumed for topic assignment and likelihood function is computed to find topic distribution of a query. Topic distribution of the Gaussians with respect to time period is defined with Topic Proportion Vector. Least Absolute Shrinkage and Selection Operator (LASSO) are used to predict total number of conversion.

9.2 Related Works

In this section, various probabilistic models for topic modelling are reviewed. Documents topics can be discovered by clustering similar documents. A non-parametric clustering algorithm is proposed based on local shrinking in that the number of convergent points are used as the number of clusters [3]. The idea is to transform data points towards denser regions. The Shrinking process is based on K-nearest neighbour method. Value of K is selected automatically based on the optimized value of Silhouette and CH index. The non-parametric clustering algorithm outperforms traditional algorithms for all given datasets even when they are provided with true number of clusters as parameters. A clustering algorithm that can identify categories from the query keywords and assign advertisement to them is proposed [4]. This algorithm is based on Bernoulli distribution. Beta priors are used to maintain discrete probability distribution to assign clusters.

Document Influence Model (DIM) model is proposed based on Dynamic Topic Model (DTM) to find topics [5] over the time. Experiments are conducted on cited documents. This model computes the sequences of topics, posterior distribution of the latent variables and the per-document influence values. It shows how past articles decide the varying influence on future articles. Article's influence value is the hidden variable and the influential articles are identified by posterior inference.

Latent Dirichlet Allocation (LDA) model is widely used to assign topics to the documents and it is an unsupervised model in which words in the documents are modelled. A supervised Latent Dirichlet Allocation (sLDA) model [6] is introduced that accepts various response types. The response variable can be movie rating, count of number of users who selected an article important, document category. Hidden topics are discovered by combining document and response and later used to predict the response variables. The sLDA model is limited to one variable association with document. A generative labelled LDA (L-LDA) model [7] is presented with multi-label supervision. Each label is associated with one topic directly. This model is a combination of supervised LDA and mixture model of Multinomial Naive Bayes. This L-LDA model is used to assign topics to Twitter profiles correctly and find similarity of profile pairs [8]. It also re-ranks Twitter blogs and suggests new users to follow. This L-LDA performs similar to Support Vector Machine and it outperforms when training data is limited.

LDA models do not find the correlations between topics. Li et al. [9] introduced Pachinko Allocation Model (PAM) which captures the relation between topics by using Directed Acyclic Graph (DAG). Each leaf in the DAG represents a word in a vocabulary and each internal node represents a relation between either words or topics. PAM with DAG does not represent the word's topical distribution that is present in multiple topics. Hierarchical PAM (hPAM) [10] arranges topics in hierarchies. This model combines the hierarchical nature of hLDA with the topic mixing abilities of PAM. In hPAM, each node is associated with distribution over the vocabulary. The resulting model is effective at discovering mixtures of topic hierarchies.

A hierarchical algorithm which represents documents as a hierarchy of latent topics computed with Dirichlet process [11]. It is based on Bayesian priors and hierarchy of topics is derived without estimating depth of the hierarchy and branching at each level. The internal nodes represent words and topics probability distribution and vocabulary clustering is performed. Leaf nodes represent word distribution in a corpus hierarchical topic clustering. This model does not restrict on topic usage, allows multiple inheritance between topics and internal nodes are modelled as subtopics and words distribution. Method proposed in [12–14] can be used for prediction.

Ravi et al. [15] proposed a probabilistic method that generates bid phrases for online advertising. This model was first trained on search query log and generates well-formed bid phrases. Next, it generates novel bid phrases from Web pages and corpus of bid phrases. Fujita et al. [16] proposed a method that generates shop-specific *ad* listing by incorporating promotional text data for restaurant domain. Experimental result shows that the click through rate is higher for automatically generated *ad* than template-based *ad*.

9.3 PCAEM Model and Algorithm

9.3.1 Problem Definition

Given a search log l and number of epoch e, the objective is to predict the total conversion in advertising.

9.3.2 Prediction Conversion in Advertising Using Expectation–Maximization Model

Predicting Conversion in Advertising using Expectation–Maximization (PCAEM) model provides the advertisers with information on the effectiveness of their advertising campaigns. This effectiveness can be measured by estimating the number of conversions with respect to time period. This framework helps the advertisers in making business decisions. Expectation–Maximization (EM) that is used to construct the proposed model and Latent Dirichlet Allocation (LDA) which is used for comparison with the proposed model are explained.

Expectation–Maximization (EM)
Expectation–maximization is an iterative process used to compute the maximum likelihood of parameters in a statistical model, especially in models where some data is unobserved [17–19]. It is used as a clustering technique. Each cluster can be mathematically represented as a probability distribution, characterized by its mean and variance. EM assumes that the data is generated by a mixture of underlying probability distribution. It assigns each object to a cluster depending on the likelihood of its membership. There are no restrictive boundaries between the clusters, i.e. an object can belong to multiple clusters with same or different probability of membership.

EM algorithm comprises two distinct steps; the Expectation step ($E - Step$) and the Maximization step ($M - Step$). In the $E - step$, the estimated parameters are used to compute the likelihood of the model. In the $M - step$, the likelihood computed in the previous step is used to determine new values for the model parameters. The algorithm converges when there is no significant change in the model.

Consider two coins A and B are tossed $five$ times, the probability p of head coming up on these tosses is known as $P = p_1, p_2 \dots p_5$. The goal is to estimate the identity of the coin for these tosses as $Q = q_1, q_2 \dots q_5$ where q_i is the identity of the coin for the ith toss. Here Q is the latent or hidden variable.

For this problem, computing the proportion of heads for each coin is not a viable option. However, the parameter estimation with incomplete data can be transformed into maximum likelihood estimation with complete data using EM. The following steps are involved:

1. Initial values for P_A and P_B are selected randomly.
2. For each of the five flips, estimate the likelihood of the flip being made using coin A or B.
3. Assuming these completions to be correct, calculate the new values for parameters P_A and P_B.
4. Repeat Steps 2 and 3 until the difference between the previous and the current estimates is negligible. This defines the convergence point of the problem.

Latent Dirichlet Allocation (LDA)
Latent Dirichlet Allocation (LDA) [20] is a common method for topic modelling. It is a generative model which separates words into different topics from a *corpus*. LDA assumes that each document is a mixture of different topics, where each topic has some particular probability of generating a particular word. A document is assumed to be a *bag of words* pulled from a distribution selected by a Dirichlet process.

Consider a corpus C which is a collection of documents D. The goal is to discover K topics from C, topic distribution of D and words associated with each topic. LDA performs the following step to achieve this goal:

1. Each word W in a document D is randomly assigned to a topic T.
2. This random assignment gives a basic structure of the goal, i.e. initial topic distribution of the document and the word distribution for the topics.
3. For each word W in D, the conditional probabilities $P(T/D)$ and $P(W/T)$ are estimated. $P(T/D)$ represents the probability of the words in D which are assigned to topic T. $P(W/T)$ is the probability of W assigned to topic T over all documents in the corpus.
4. Each word is re-sampled and it is assumed that the topic assignment for the current word is incorrect. Word W is then assigned a new topic by computing $P(T/D)*p(W/T)$ that gives the probability of topic T containing word W.
5. The previous step is iterated until a steady state is reached. The assignments at this state are used to estimate the topic distribution of the documents and the word distribution of the topics.

The PCAEM model has the following steps:

Step 1: Building a Vocabulary
In this step, the queries from the query log are considered to build a vocabulary. Stopwords *a, an, the, for, etc.* are used as grammar constructs and don't convey any meaning to the query. Hence, they are removed from the queries to obtain the words that convey the requirements of the users. The vocabulary V is built from the remaining words in the queries after removal of stopwords and the frequency of occurrence of those words is computed. Consider a query *create an android app for generating recipe*, in which the words *an* and for are removed. After preprocessing, the query now becomes *create android app generating recipe*.

Step 2: Gaussian Assumption for Defining Topics
A query is a mixture of various topics. Hence it is necessary to determine the latent topics in a query. Consider the query *create an android app for generating recipe*, it comprises words related to both *technology* as well as *food*. In EM, topics are represented as Gaussians *Gauss[K]*, where K is the number of topics. A Gaussian is characterized by its mean and variance. Gaussians contain all the words in the vocabulary, in which each word is randomly assigned a topic between 1 to K. The initial probabilities of the words belonging to topic i in Gaussians are equal, where $i \in 1, \ldots k$. The initial mean of a Gaussian and covariance between the Gaussians is computed as shown in Function 9.1.

Function 9.1: Initial Mean and Covariance Generation

Function: InitialTopic
Data: Consider Vocabulary V and Number of Topics K. Initial Mean and Covariance Matrix are generated.

Let n = number of Gaussians = number of topics
$mean[n][K]$ = mean of each topic in K for n Gaussian
$covariance[n][n]$ = covariance between Gaussians
$Gauss[n]$ = n Gaussians
V_{size} = vocabulary size
for i = *1 to n* **do**
    $Gauss[i]$ = All vocabulary words
    for j = *1 to* V_{size} **do**
        $Gauss[i].word[j]$ = random topic assignment between 1 to K
    for L = *1 to K* **do**
        Let Num_{word} = Number of words in Gaussian i with topic K
        Let Tot_{word} = Total number of words in Gaussian i
        Calculate mean for topic L in $Gauss[i] = Mean[i][L] = \frac{Num_{word}}{Tot_{word}}$
for i = *1 to n* **do**
    for j = *1 to n* **do**
        Calculate covariance(Gauss[i],Gauss[j]) using Eq. 9.1

$$covariance_c(i, j) = [(Prob_{Gauss_i}) - (mean[i][topic(word_c, i))]* \\ [(Prob_{Gauss_j}) - (mean[j][topic(word_c, j))] \tag{9.1}$$

Here, $Prob_{Gauss_i}$ is the probability of $word_c$ in Gauss[i], $Prob_{Gauss_j}$ is the probability of $word_c$ in Gauss[j], $topic(word_c,i)$ is the topic $word_c$ in Gaussian i and $mean[i][topic(word_c,i]$ is the mean of $topic(word_c,i)$ in Gaussian i.

Step 3: Compute Likelihood to find Topic Distribution of a Query

In the previous step, the topics defined in the Gaussians are estimated. Using these Gaussians, the topic distribution of the queries is computed. All the queries are assumed to be stationary points in a space. This space also contains the Gaussians with equal probability $P(c)$. As the Gaussians move around and take shape, the conditional probability of occurrence $P(qi/c)$ of a query qi in the Gaussian c changes. A query qi can belong to more than one topic. Consider the query *create an android app for generating recipe*, the words $android$ and app come from the $technology$ topic and the word $recipe$ comes from the $food$ topic, hence this query occurs in the $technology$ topic with probability (2/5) and in the $food$ topic with probability (1/5). The likelihood of a query occurring in a topic is computed as shown in Function 9.2.

Function 9.2: Query Topic Likelihood

Function: QueryTopicLikelihood

Data: Consider Mean matrix $mean[n][k]$, Covariance matrix $cov[n][n]$, Number of queries Q_{num} and n is the number of Gaussians. Conditional Probability of Topic C for given Query q PCQ, Conditional Probability of Query q for given Topic C PQC and Probability of Gaussian $P(C)$ are generated.

for i = *1 to* Q_{num} **do**
    for c = *1 to* K **do**
        Compute $PQC(i,c)$ using Eq. 9.2

for c = *1 to* K **do**
    for i = *1 to* Q_{num} **do**
        Compute $PCQ(c,i)$ using Eq. 9.3

for c = *1 to* n **do**
    Compute $P(c)$ using Eq. 9.4

$$PQC(i,c) = \frac{1}{\sqrt{2\pi \mid cov_c \mid}} \left\{ \frac{-1}{2} \left[\sum_{K=1}^{n} \sum_{j=1}^{n} ((q_{i,j} - mean(c,j)) * (q_{i,k} - mean(c,k))) * \left(cov_c\right)^{-1}_{j,k} \right] \right\} \tag{9.2}$$

Here, $\mid cov_c \mid$ is the determinant of $covariance_c$.

$$PCQ(c,i) = \frac{PQC(i,c) * P(c)}{\sum_{c'=1}^{n} PQC(i,c') * P(c')} \tag{9.3}$$

$$P(c) = \frac{1}{n} \sum_{j=1}^{Q_{num}} PCQ(c,j) \tag{9.4}$$

Step 4: Compute New Definition of Topics

The initial mean and covariance of the Gaussians give a basic structure to the topic. The conditional probabilities estimated in the previous step influence the mean and covariance of the Gaussian such that the mean of the topics in the Gaussian changes. Hence, the Gaussians need to be recomputed with respect to the conditional probabilities. The mean and covariance of the Gaussians is computed using Eqs. 9.5 and 9.6.

$$mean(c, j) = \sum_{i=1}^{Q_{num}} \left[\frac{PCQ(c, i)}{cP(c)} \right] q_{i,j} \tag{9.5}$$

$$covariance_c(j, k) = \sum_{j=1}^{Q_{num}} \left(\frac{PCQ(c, i)}{nP(c)} \right) * \left(q_{i,j} - mean(c, j) \right) * \left(q_{j,k} - mean(c, k) \right) \tag{9.6}$$

Step 5: Test for Convergence of the Model
Convergence signifies a stable state of the model. After a certain number of iterations, the Gaussians gain a definite shape and position in space where the change in the conditional probabilities is negligible. Hence, Steps 3 and 4 need to be iterated until the model converges.

Step 6: Compute Topic Proportion Vector
Topic proportion vector TPV represents the topic distribution of the Gaussians in the model. It is obtained from the mean computed at the convergence. As this model predicts the total number of conversions for advertisement campaign in a particular time period which is called an *epoch*. TPV for each epoch is defined as $TPV[e][i]$, where e and i represent epoch and topic, respectively. TPV for an epoch represents the weight of the topics in that epoch.

Step 7: Conversion Prediction
A conversion occurs when a user clicks on the advertisement and performs some action that gives benefit to the advertiser. This type of user click is defined as Cost Per Click (CPC). In order to predict the total number of conversions, the model is trained using the L_1 prior also known as $LASSO$ (Least Absolute Shrinkage and Selection Operator) [21]. The conversion for each epoch is estimated using Eq. 9.7.

$$conversion = \frac{1}{2 * T} * \| CPC - TPV * R \|_2^2 + \alpha + \| R \|_1 \tag{9.7}$$

Here, α is the hyper parameter, R is the Regression and CPC is the cost per click.

9.3.3 PCAEM Algorithm

Predicting Conversion in Advertising using Expectation–Maximization (PCAEM) algorithm as shown in Algorithm 9.1 has two phases: Preprocessing and Prediction. Preprocessing involves building vocabulary, Gaussian assumption, computing likelihood for topic distribution of queries, re-estimation of the Gaussians and constructing topic proposition matrix. Prediction involves LASSO-based conversion prediction.

Algorithm 9.1: Predicting Conversion in Advertising using Expectation Maximization

Input : Query log l, Number of epoch N_e
Output: Conversion prediction cp

```
begin
    Pre-processing :
    Build vocabulary V by considering queries from log l
    for m = 1 to N_e do
        Assume K Gaussians to represent topics as explained in step 2.
        while model not converged do
            Estimate the likelihood to find topic distribution of each query as per step 3.
            Re-estimate Gaussians to find new definition of the topics as per step 4.
        for n = 1 to K do
            Construct Topic Proportion Vector TPV[m][n] using the final Gaussian
            estimation after convergence.
    Prediction :
    Predict conversions using LASSO method as per step 7
```

9.4 Experiments

9.4.1 Data Collection

In this experiment, query data is used in [22] which is collected from the U.K and the U.S market over 34 weeks is considered. It consists of various queries ranging over multiple topics. This dataset consists of the following information: Campaign Name, Country, Geo Targeting, ID, Budget, Ad Group, Max. CPC, Keyword, Language, Keyword Type, Average position, Average CPC, Clicks, CTR, Cost and Impressions. Here, Click Through Rate (CTR) is the ratio of how often user clicks an advertisement that appears as a relevant result for the query. It evaluates the competence of the keywords and the advertisement. An impression signifies the relevance of an advertisement and is incremented every time it appears as a result of a searched query. In the proposed model, keywords are used to generate the Topic Proportion Matrix and CPC influence the conversion. Hence, Keyword and Average CPC is used in this experiment.

9.4.2 Experimental Setup

The proposed model PCAEM uses probabilistic estimations for topic modelling. LDA also works with the same principle, hence PCAEM is compared with TopicMachine [22].

The setup of PCAEM is as follows: Vocabulary is constructed from the keywords. The words in the vocabulary are ranked based on their frequency of occurrence. Words with low frequency do not influence the topic proportion. Hence, the size of the vocabulary is restricted to 500. Experiments are conducted with varying topic number $K = 5, 6, 7, 8, 9, 10, 11$. The initial probability of each topic is considered equal. Using these assumptions the topic distribution of each query is estimated. The model is iterated 10 times for convergence. It is observed that PCAEM model converges at the fifth iteration. Hence, the number of conversions is estimated after the fifth iteration. The data is collected for 34 weeks, hence the number of epochs are 34 as 1 epoch is for 1 week. The experiment is conducted by varying epoch size to 15, 20, 25, 30 and 34. Topic proportion vector is constructed for each epoch and used for conversion prediction.

The setup for TopicMachine is as follows: Vocabulary is constructed in a similar way as PCAEM. Initially, words in the queries are randomly assigned to topics. These assignments are used to approximate hidden variables using variational approximation iteratively. Variation threshold is set to define a convergence point for the model. Experiments are conducted by varying the threshold parameter to 0.1, 0.01 and 0.001. It is observed that definite topics are obtained when threshold is set to 0.001. Topic proportion vector is constructed for each epoch and used for conversion prediction.

9.4.3 Performance Metrics

The performance metrics used for comparison are Root Mean Squared Error (RMSE), Mean Absolute Error (MAE) and Predictive R^2, i.e. the coefficient of determination. The actual value for each epoch corresponds to the average CPC collected from the dataset. Consider E is the total number of epochs, dataset is partitioned into training set T and testing set Te, then $offset = EXTe/100$. The predicted and actual values are denoted by p and a, respectively.

RMSE is the root of the square of the difference between the values predicted by the model and the actual values as shown in Eq. 9.8. MAE is the average of the absolute value of the difference between the values predicted by the model and the actual values as shown in Eq. 9.9. Coefficient of determination is computed as shown in Eq. 9.10. It measures the correctness of the model based on the proportion of deviation of predicted value from actual value.

$$RMSE = \sqrt{\sum_{i=E-offset+1}^{E} (a_i - p_i)^2} \tag{9.8}$$

$$MAE = \frac{1}{N} \sum_{i=E-offset+1}^{E} | p_i - a_i | \tag{9.9}$$

$$R^2 = 1 - \frac{\sum_{i=E-offset+1}^{E} (a_i - p_i)^2}{\sum_{i=E-offset+1}^{E} (a_i - mean(p_i))^2} \tag{9.10}$$

9.4.4 Performance Evaluation

In this section, experimental results are presented and discussed. Performance metrics are used to compare the results of the proposed model PCAEM and TopicMachine [22]. Experiments are conducted by varying training dataset to 70, 75, 80, 85, 90% and testing dataset to 30, 25, 20, 15, 10%. Experiments have been conducted on 4 GB memory and Intel(R) Core(TM) i5-5200U CPU @ 2.20 GHz processor. Dataset used in the experiments for PCAEM and TopicMachine is the same as discussed in data collection. The performance is evaluated by computing the average of 20 independent runs.

RMSE, MAE and R^2 by Varying Number of Topic K

In order to measure the effect of the number of topics K, all the performance metrics are computed to measure the performance of PCAEM and TopicMachine. The value of K varies from 5 to 11. Table 9.1 shows the comparison of RMSE, MAE and R^2 values for both the methods by setting vocabulary size to 500, α to 0.01, epoch to 34 and training dataset to 85%. It is observed from the table that the number of topics does not affect much on the performance of PCAEM and TopicMachine. It is also observed from the table that the RMSE and MAE values are lesser in PCAEM than TopicMachine, hence the performance of PCAEM is better than TopicMachine.

RMSE, MAE and R^2 by Varying Training Dataset

In order to analyse the performance of models with available training dataset, RMSE, MAE and R^2 is computed by increasing training dataset. Table 9.2 shows the comparison of RMSE, MAE and R^2 values for both the methods by setting vocabulary size to 500, α to 0.01, Topic number K to 10 and epoch to 34. It is observed from Table 9.2 that as the training dataset increases the RMSE decreases and MAE increases marginally.

RMSE, MAE and R^2 by Varying Epoch Size

In order to study the consequence of the size of the dataset on the performance of PCAEM and TopicMachine, RMSE, MAE and R^2 is computed by varying the epoch

Table 9.1 RMSE, MAE and R^2 by varying the number of topics K for PCAEM and TopicMachine. The vocabulary size, α, epoch and training dataset are set to 500, 0.01, 34 and 85%, respectively

	PCAEM method			TopicMachine		
Topic	RMSE	MAE	R^2	RMSE	MAE	R^2
5	8.6142	4.3039	−18.6755	8.6898	4.3416	−19.0221
6	8.6071	4.3004	−18.6431	8.6820	4.3377	−18.9863
7	8.6054	4.2995	−18.6354	8.6839	4.3387	−18.9949
8	8.6040	4.2988	−18.6286	8.6838	4.3386	−18.9948
9	8.6034	4.2985	−18.626	8.6794	4.3364	−18.9743
10	8.6019	4.2978	−18.6193	8.6795	4.3365	−18.9751
11	8.6015	4.2975	−18.6172	8.6706	4.3321	−18.934

Table 9.2 RMSE, MAE and R^2 by varying training dataset for PCAEM and TopicMachine. The vocabulary size, α, Topic Number K and epoch are set to 500, 0.01, 10 and 34, respectively

	PCAEM method			TopicMachine		
Training dataset	RMSE	MAE	R^2	RMSE	MAE	R^2
70	12.2351	4.0718	−6.1212	12.3420	4.1073	−6.2462
75	10.9441	4.1300	−7.7055	11.0403	4.1662	−7.8592
80	9.4557	4.2235	−11.8799	9.5403	4.2612	−12.1114
85	8.6019	4.2978	−18.6193	8.6795	4.3365	−18.9751
90	6.2202	4.3970	−40.7745	6.2770	4.4370	−41.5397

Table 9.3 RMSE, MAE and R^2 by varying epoch size for PCAEM and TopicMachine. The vocabulary size, α, Topic Number K and training dataset are set to 500, 0.01, 10 and 85% respectively

	PCAEM method			TopicMachine		
Epoch	RMSE	MAE	R^2	RMSE	MAE	R^2
15	3.4425	3.4425	0.8266	3.5394	3.5394	0.8168
20	6.7551	4.7729	−19.9357	6.8263	4.8230	−20.379
25	5.0849	3.5898	−0.4554	5.1259	3.6187	−0.4790
30	6.7661	3.9063	−3.0550	6.6234	3.9394	−3.1240
34	8.6019	4.2978	−18.6193	8.6795	4.3365	−18.9751

size. Table 9.3 shows the comparison of RMSE, MAE and R^2 values for both the methods by setting vocabulary size to 500, α to 0.01, Topic number K to 10 and training dataset to 85%. There is not much difference observed by varying size of the dataset.

Table 9.4 RMSE, MAE and R^2 by varying hyper parameter α for PCAEM and TopicMachine. The vocabulary size, Topic Number K, epoch and training dataset are set to 500, 10, 34 and 90% respectively

	PCAEM method			TopicMachine		
α	RMSE	MAE	R^2	RMSE	MAE	R^2
100	80.9580	57.2459	−7075.36	6.2769	4.4370	−41.5388
50	37.3657	26.4213	−1506.43	6.2778	4.4377	−41.5518
30	19.9290	14.0915	−427.807	6.2771	4.4371	41.541
20	11.2110	7.9266	−134.7	6.2771	4.4372	−41.542
10	**2.4962**	**1.7617**	**−5.7279**	6.2773	4.4373	−41.5442
1	5.3574	3.7867	−29.9892	6.2772	4.4372	−41.5429
0.1	6.1418	4.3415	−39.7275	6.2778	4.4376	−41.5508
0.01	6.2202	4.3970	−40.7745	6.2778	4.4376	−41.551
0.001	6.2281	4.4025	−40.8799	6.2774	4.4374	−41.5461

PCAEM and TopicMachine Sensitivity to Hyper Parameter α

The hyper parameter α is used for prediction conversion, it is necessary to study the effect of changes in α affect the quality of model. RMSE, MAE and R^2 are computed by varying the α value for both the models. It is observed from Table 9.2 that RMSE values decreases when training dataset increases, hence, experiments are conducted by setting training dataset 90 and 85%. Table 9.4 shows the comparison of RMSE, MAE and R^2 values for both the methods by setting vocabulary size to 500, Topic number K to 10, epoch to 34 and training dataset to 90%. Table 9.5 shows the comparison of RMSE, MAE and R^2 values for both the methods by setting vocabulary size to 500, Topic number K to 10, epoch to 34 and training dataset to 85%. It is observed from Tables 9.4 and 9.5 that PCAEM model is sensitive to hyper parameter α, but TopicMachine is not. PCAEM model is giving best result when α is set to 10.

It is observed from Tables 9.4 and 9.5 that PCAEM model is giving best results when α is set to 10. Hence, RMSE and MAE is computed again by varying training dataset, by varying the number of the topic K and by varying the *epoch* size by setting α value to 10. Figures 9.1 and 9.2 show the comparison of RMSE and MAE by Varying Training Dataset when α is set to 10, respectively. It is observed that the average value of RMSE is 4.4572 and 9.5752 of PCAEM and TopicMachine, respectively. The average value of MAE is 1.9346 and 4.2614 of PCAEM and TopicMachine, respectively.

Figures 9.3 and 9.4 show the comparison of RMSE and MAE by the varying number of topics K when α is set to 10, respectively. It is observed that the average value of RMSE is 3.2013 and 8.6815 of PCAEM and TopicMachine, respectively. The average value of MAE is 1.5861 and 4.3375 of PCAEM and TopicMachine, respectively.

Table 9.5 RMSE, MAE and R^2 by varying hyper parameter α for PCAEM and TopicMachine. The vocabulary size, Topic Number K, epoch and training dataset are set to 500, 10, 34 and 85%, respectively

	PCAEM method			TopicMachine		
α	RMSE	MAE	R^2	RMSE	MAE	R^2
100	114.6907	57.3451	−3486.75	8.6793	4.3364	−18.974
50	53.0421	26.5205	−744.988	8.6791	4.3362	−18.9728
30	28.3834	14.1907	−212.609	8.6790	4.3362	−18.9723
20	16.0551	8.0258	−67.3464	8.6797	4.3366	−18.976
10	**3.7365**	**0.8609**	**−2.7018**	8.6794	4.3364	−18.9746
1	7.3823	3.6874	−13.4503	8.6794	4.3365	−18.9746
0.1	8.4910	4.2423	−18.1167	8.6797	4.3366	−18.9758
0.01	8.6019	4.2978	−18.6193	8.6793	4.3364	−18.9738
0.001	8.6130	4.3033	−18.6699	8.6796	4.3365	−18.9755

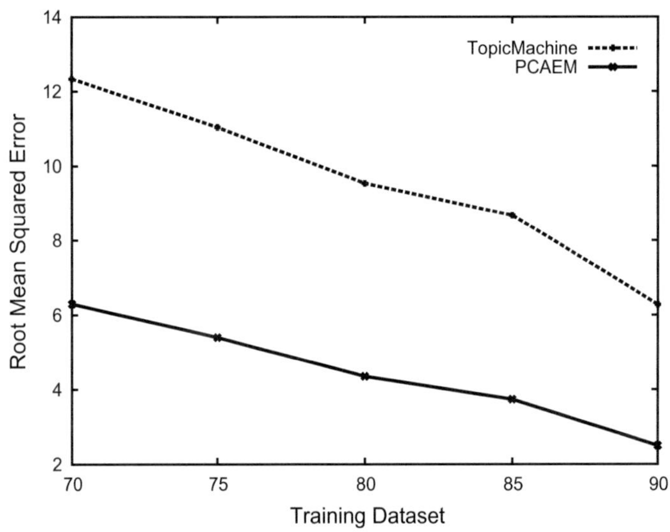

Fig. 9.1 RMSE by varying training dataset when $\alpha = 10$

Figures 9.5 and 9.6 show the comparison of RMSE and MAE by varying the *epoch* size when α is set to 10 respectively. It is observed from that the average value of RMSE 3.1952 and 6.1987 of PCAEM and TopicMachine respectively. The average value of MAE is 2.1568 and 4.0513 of PCAEM and TopicMachine respectively.

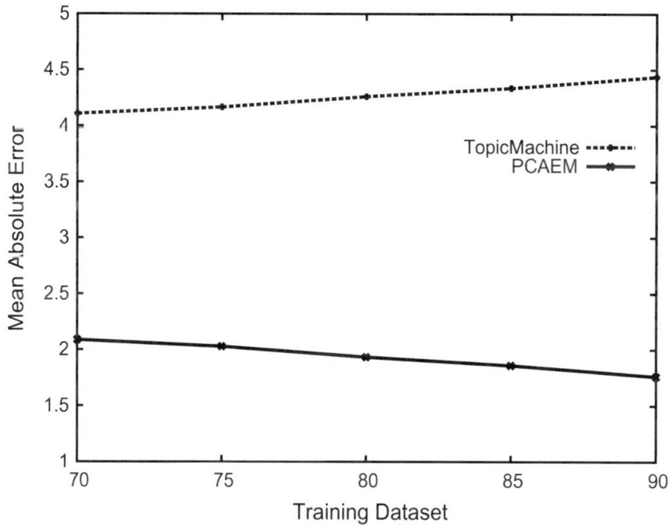

Fig. 9.2 MAE by varying training dataset when $\alpha = 10$

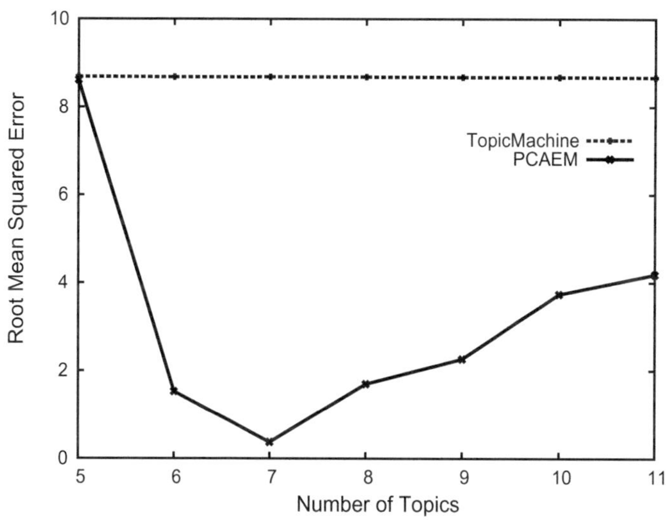

Fig. 9.3 RMSE by varying number of topics when $\alpha = 10$

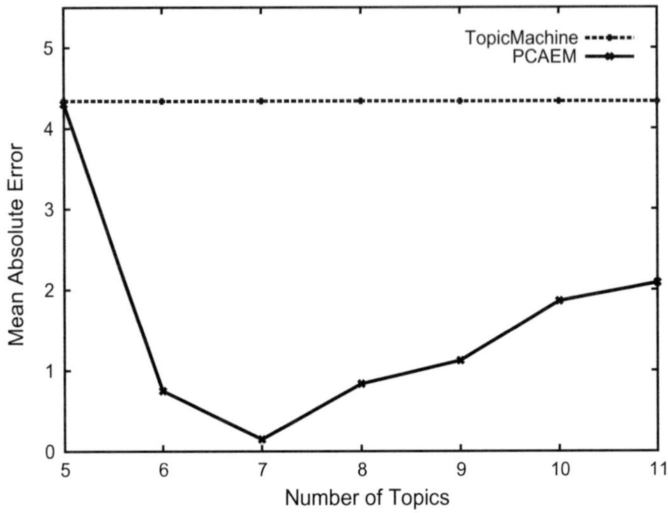

Fig. 9.4 MAE by varying number of topics when $\alpha = 10$

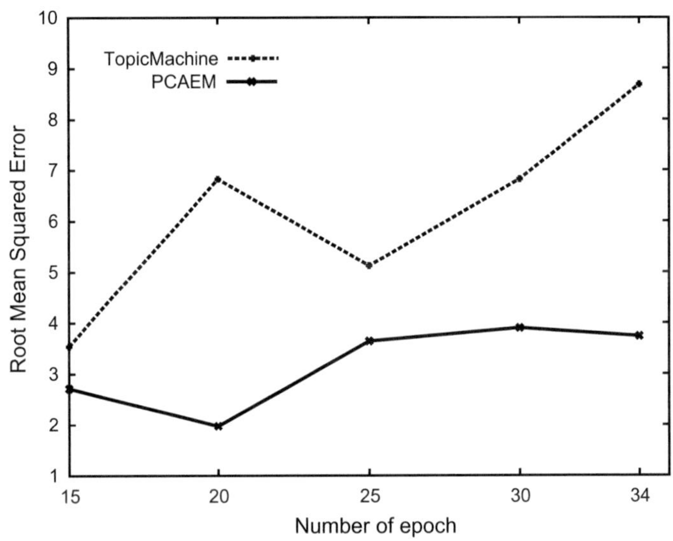

Fig. 9.5 RMSE by varying *epoch* when $\alpha = 10$

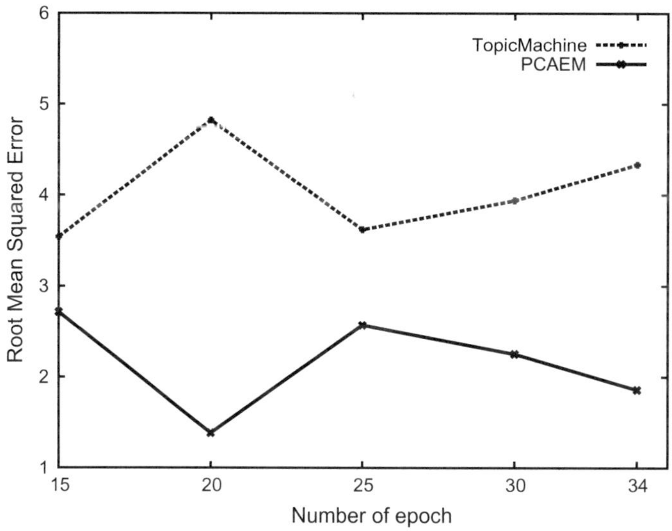

Fig. 9.6 MAE by varying *epoch* when $\alpha = 10$

9.5 Summary

In this chapter, we present predicting conversion in advertising using expectation–maximization [PCAEM] model to understand advertising campaigns' effectiveness to the advertisers over the time. Search query log is used to build vocabulary. Expectation–Maximization method is used to find the hidden topics and topic distribution on search terms. Least Absolute Shrinkage and Selection Operator (LASSO) is used to predict total number of conversion. Experiments are performed on query data used in [22] which is collected from the U.K and the U.S market over 34 weeks. Root Mean Squared Error (RMSE), Mean Absolute Error (MAE) and Predictive R^2 are used as performance metrics. Experiment results are compared with TopicMachine [22]. The proposed method outperforms TopicMachine [22] by reducing RMSE and MAE. The PCAEM model is sensitive to hyper parameter used for prediction conversion while TopicMachine is not.

References

1. B. Ribeiro-Neto, M. Cristo, P.B. Golgher, E. Silva de Moura, Impedance coupling in content-targeted advertising, in *The Proceedings of the 28th Annual International ACM SIGIR Conference on Research and Development in Information Retrieval* (2005), pp. 496–503
2. P. Rusmevichientong, D.P. Williamson, An adaptive algorithm for selecting profitable keywords for search-based advertising services, in *Proceedings of the 7th ACM Conference on Electronic Commerce* (2006), pp. 260–269

3. X. Wang, W. Qiu, R.H. Zamar, An iterative non-parametric clustering algorithm based on local shrinking. Comput. Stat. Data Anal. **52**, 286–298 (2007)
4. A. Schwaighofer, J.Q. Candela, T. Borchert, T. Graepel, R. Herbrich, Scalable clustering and keyword suggestion for online advertisements, in *Proceedings of the 3rd International Workshop on Data Mining and Audience Intelligence for Advertising* (2009), pp. 27–36
5. S. Gerrish, D.M. Blei, A language-based approach to measuring scholarly impact, in *Proceedings of the 27th International Conference on Machine Learning (ICML-10)* (2010), pp. 375–382
6. J.D. Mcauliffe, D.M. Blei, Supervised topic models, in *Advances in Neural Information Processing Systems* (2008), pp. 121–128
7. D. Ramage, D. Hall, R. Nallapati, C.D. Manning, Labeled LDA: a supervised topic model for credit attribution in multi-labeled corpora, in *Proceedings of the 2009 Conference on Empirical Methods in Natural Language Processing: Volume 1* (2009), pp. 248–256
8. D. Quercia, H. Askham, J. Crowcroft, TweetLDA: supervised topic classification and link prediction in Twitter, in *Proceedings of the 4th Annual ACM Web Science Conference* (2012), pp. 247–250
9. W. Li, A. McCallum, Pachinko allocation: DAG-structured mixture models of topic correlations, in *Proceedings of the 23rd International Conference on Machine Learning* (2006), pp. 577–584
10. D. Mimno, W. Li, A. McCallum, Mixtures of hierarchical topics with Pachinko allocation, in *Proceedings of the 24th International Conference on Machine Learning* (2007), pp. 633–640
11. E. Zavitsanos, G. Paliouras, G.A. Vouros, Non-parametric estimation of topic hierarchies from texts with hierarchical Dirichlet processes. J. Mach. Learn. Res. **12**, 2749–2775 (2011)
12. K.G. Srinivasa, K.R. Venugopal, L.M. Patnaik, An efficient fuzzy based neuro-genetic algorithm for stock market prediction. Int. J. Hybrid Intell. Syst. **3**(2), 63–81 (2006)
13. K.C. Srikantaiah, M.S. Roopa, N. Krishna Kumar, K.R. Venugopal, L.M. Patnaik, Automatic discovery and ranking of synonyms for search keywords in the web. Int. J. Web Sci. **2**(4), 218–236 (2014)
14. K.C. Srikantaiah, N. Krishna Kumar, K.R. Venugopal, L.M. Patnaik, Web caching and prefetching with cyclic model analysis of web object sequences. Int. J. Knowl. Web Intell. 2, **5**(1), 76–103 (2014)
15. S. Ravi, A. Broder, E. Gabrilovich, V. Josifovski, S. Pandey, B. Pang, Automatic generation of bid phrases for online advertising, in *Proceedings of the 3rd ACM International Conference on Web Search and Data mining* (2010), pp. 341–350
16. A. Fujita, K. Ikushima, S. Sato, R. Kamite, K. Ishiyama, O. Tamachi, Automatic generation of listing ads by reusing promotional texts, in *Proceedings of the 12th International Conference on Electronic Commerce: Roadmap for the Future of Electronic Business* (2010), pp. 179–188
17. A.P. Dempster, N.M. Laird, D.B. Rubin, Maximum likelihood from incomplete data via the EM algorithm. J. R. Stat. Soc. Ser. B (Methodol.), 1–38 (1977)
18. J.A. Bilmes et al., A gentle tutorial of the EM algorithm and its application to parameter estimation for Gaussian mixture and hidden Markov models. J. Int. Comput. Sci. Inst. **4**, 1–15 (1998)
19. K.M. Todd, The expectation-maximization algorithm. IEEE Signal Process. Mag. **13**(6), 47–60 (1996)
20. D.M. Blei, A.Y. Ng, M.I. Jordan, Latent Dirichlet allocation. J. Mach. Learn. Res. **3**, 993–1022 (2003)
21. R. Tibshirani, Regression shrinkage and selection via the lasso. J. R. Stat. Society. Ser. B (Methodol.), 267–288 (1996)
22. A. Bulut, TopicMachine: conversion prediction in search advertising using latent topic models. IEEE Trans. Knowl. Data Eng. **26**(11), 2846–2858 (2014)

Further Reading

1. K.C. Srikantaiah et al., PCF-engine: a fact based search engine, in *Computer Networks and Intelligent Computing. ICIP 2011*, vol. 157, ed. by K.R. Venugopal, L.M. Patnaik (Springer, Berlin, Heidelberg, 2011)
2. D. Sejal et al., QRGQR: query relevance graph for query recommendation, in *IEEE Region 10 Symposium*. (Ahmedabad, 2015), pp. 78–81
3. D. Sejal, V. Rashmi, K.R. Venugopal et al., Image recommendation based on keyword relevance using absorbing Markov chain and image features. Int. J. Multimed. Inf. Retr. **5**, 185–199 (2016). https://doi.org/10.1007/s13735-016-0104-9
4. D. Sejal, D. Abhishek, K.R. Venugopal et al., IR_URFS_VF: image recommendation with user relevance feedback session and visual features in vertical image search. Int. J. Multimed. Inf. Retr. **5**, 255–264 (2016). https://doi.org/10.1007/s13735-016-0111-x
5. D. Sejal, T. Ganeshsingh, K.R. Venugopal et al. ACSIR: ANOVA cosine similarity image recommendation in vertical search. Int. J. Multimed. Inf. Retr. **6**, 143–154 (2017). https://doi.org/10.1007/s13735-017-0124-0

K. R. Venugopal et al., *Web Recommendations Systems*,
https://doi.org/10.1007/978-981-15-2513-1

Index

K. R. Venugopal et al., *Web Recommendations Systems*,
https://doi.org/10.1007/978-981-15-2513-1

MIX
Papier aus verantwortungsvollen Quellen
Paper from responsible sources
FSC® C105338

If you have any concerns about our products,
you can contact us on
ProductSafety@springernature.com

In case Publisher is established outside the EU,
the EU authorized representative is:
Springer Nature Customer Service Center GmbH
Europaplatz 3, 69115 Heidelberg, Germany

Printed by Libri Plureos GmbH
in Hamburg, Germany